TELEPHONES:

Antique to Modern

TELEPHONES:

Antique
to Modern

Kate E. Dooner

4880 Lower Valley Road, Atglen, PA 19310 USA

* Prices provided by and further information can be obtained from: 20th Century Vintage Telephones, 2780 Northbrook Place, Boulder , CO 80304.

To my parents, without whom I
would not be where I am today.

One half title page photo:
Ness AutomaticTelephone system manufactured by the Holzer-Cabot Electric Co., c
1900s. Wooden potbelly, OST receiver.
12.5". *Courtesy of Private Collection.*

Published by Schiffer Publishing, Ltd.
4880 Lower Valley Road
Atglen, PA 19310
Phone: (610) 593-1777 Fax: (610) 593-2002
E-mail: schifferbk@aol.com
Please write for a free catalog.
This book may be purchased from the publisher.
Please include $3.95 for shipping.
Try your bookstore first.

We are interested in hearing from authors
with book ideas on related subjects.

Revised price guide 1997

Copyright © 1992 by Schiffer Publishing, Ltd.
Library of Congress Catalog Number: 91-67008.

Printed in United States of America.
ISBN: 0-7643-0352-X

Contents

Acknowledgments

Rare telephone with Blake transmitter, c. 1879. Made by Charles Williams, Jr. for the National Bell Telephone Co. Walnut, 19.5'' x 8.25'' x 4''. *Courtesy of Norman M. Mulvey.*

A project such as this is in no way compiled by one person alone. I am indebted to the many people who provided support, knowledge and telephones to this book and would like to thank them all.

One of the greatest joys in carrying out this project was in meeting the people, not just the telephones. Thanks to all of the wonderful collectors who so readily lent their collections and knowledge including: Jim Aita, Robert E. Bartlett, Bob Breish, Barry Erlandson, Al Farmer, Gerald Gapa, Larry Garnatz, James A. Goodwin, G.K. Hillestad, George W. Howard, Art Hyde, Michael Irvin, Ken King, Keith E. Letsche, Odis LeVrier, Dale Martin, Beverly and Paul McFadden, Norman M. Mulvey, Zora Natanblut, Beverley and Roger Natte, Bob Newell, Burt Prall, Wesley Smith, Charles E. and Bill Stanley, Harvey Stuart, Paul Vaverchak, Dennis & Jeanne Weber, Jerry Williams, Vincent Wilson, and Nick Zervos and the private collectors who do not wish to be mentioned. Never have I met such a splendid and convivial group of people who were more help to me than could ever be imagined.

Special mention should be given to Al Farmer who spent so much time sharing his expertise to provide exact information for this book in the captions, and for proofreading the final copy. Also to George Howard and Norm Mulvey who also proofread parts of the book and gave their knowledgeable advice.

Thank you to Elliot Sivowitch from the Smithsonian Institute for spending precious time with me as I researched.

Peter and Nancy Schiffer, who believed in my abilities in such a project, thank you for the opportunity.

Thank you to my father for his excellent grammatical skills, Douglas Congdon-Martin for teaching me so much about photography, Bonnie Mann for her beneficial assistance, and Ellen J. (Sue) Taylor, who has beautifully designed this book.

Chapter One
The Invention

The idea of converting sound into electrical currents and sending them through a wire was not a wholly new idea at the time of the invention of the telephone in 1876. The concept was under research years before any type of articulated sound was transmitted.

As early as 1664, Dr. Robert Hooke, an English physicist, supported the idea of telephony by creating a device to convey sound. In 1753, Stephen Gray, another Englishman, wrote of a method for sending and receiving messages over a considerable distance. In 1837, Professor C.D. Page of Salem, Massachusetts used an iron bar to emit sound.

The first person to use the term "telephone" was Sir Charles Wheatstone, a physicist and pioneer in telegraphy who used the word in 1840. In 1851, Dr. S.D. Cushman of Racine, Wisconsin developed an "Electric Talking Box," although the instrument was never patented. In 1854, Charles Borseul, a Frenchman, was the first to suggest transmitting the human voice by electric currents, however there is no record that he ever built an instrument to work in this way. (Paper 29, p. 314) (E.H. Danner Museum of Telephony brochure).

As early as 1849, Antonio Meucci, an Italian-born candlemaker and brewer from Staten Island, claimed to have devised an instrument which would transmit sound. He made several applications for a patent in America beginning in 1871, but never received one. In the 1860s the idea was taken up by the German physicist Philipp Reis. After many trials and attempts at making his own device, Reis exhibited his telephone to the Physical Society of Frankfort in 1861. In 1864, Reis proclaimed his speaking telephone to be a true scientific invention, although not yet practical for use. Reis's memoir states, "there may probably remain much to be done toward making the telephone *of practical commercial value*. For physics, however, it has already sufficient interest in that it has opened out a new field of labor." The public, however,

demonstrated that the world was not yet prepared for such an advancement in technology. A device which could speak was considered to be something from the supernatural. (Popular Science Monthly, p.540-542).

Alexander Graham Bell (1847-1922) is legally credited with the invention of the telephone and his name is still associated with the telephone companies of America. When discussing the invention, we must not forget all those who worked before Bell and during his time, without whose insight the telephone may have been much longer in coming to be. Imagine dealing with the Gray Telephone Company or the Dolbear, or Meucci, or Reis Telephone Company, instead of the now ubiquitous name of the Bell Company, or as some refer to it today, "Ma Bell."

A.G. Bell's original telephone incorporated some of the principles used in Reis's telephone. Many reports on the subject conclude that Reis was the genius behind the general principles of any electro-magnetic receiver, including those developed by Elisha Gray, Alexander Graham Bell and Thomas Edison, three of the most notable contributors to the history of the telephone. Reis's transmitter formed the basis for the idea of moving or varying currents as did Bell's. Although a working and commercial telephone was not developed by Reis, he laid the foundations for those who followed him, and aided them on the path to invention. (Popular Science Monthly, p. 550-551).

A decade after Reis, came Alexander Graham Bell. With much guessing and irony, he worked on the harmonic telegraph which resulted in one of the greatest inventions of all time, the telephone. As Isaac Newton once stated, "No great discovery was ever made without a bold guess."

Bell was born in Edinburgh, Scotland on March 3, 1847 as Alexander Bell. (He acquired the middle name Graham at the age of eleven from a friend of the family, a Cuban planter named Alexander Graham.) Born into a family of pioneering elocutionists, his interest in speech was ingrained in him from the start. Alexander Graham Bell was an average student who experienced the common daily life of a child growing up in Scotland in the 1850s. He was an extremely hard worker, however, and his pursuits and endeavors would pay off and lead him to success at a young age.

At 15, young Bell helped his father, Melville Bell, with lectures on his invention of "Visible Speech," a written code intended to help people with the phonetics of foreign languages. However, it also proved useful in training the deaf to speak articulately. Bell's hearing was far below average in sensibility and range and his mother and future wife were deaf as well. Visible Speech provided the inspiration for George Bernard Shaw's play *Pygmalion* and in the musical version, *My Fair Lady*, where Professor Higgins represents Melville Bell.

By the age of 17, Bell, by now a full-fledged master teacher, began the research in speech and sound that led to the invention of a "harmonic telegraph." In 1865, he discovered vowel sounds to be formed in the resonance cavities of the mouth. (Beottinger, p. 41-42). During his independent research, Bell came upon the work of Hermann von Helmholtz, who authored the German book *On the Sensations of Tone*. In the book, Helmholtz describes the use of electrically driven tuning forks to produce vowel sounds. Bell was not altogether familiar with the German language, so his translation led him to believe that Helmholtz had telegraphed the vowel sounds over a wire. Although this was not really the case, it inspired Bell to turn his experiments for the first time to electric transmission. (Brooks, p. 38-39). (Bell did not realize his mistake until 1870 when he read the book in French. (Beottinger, p.45).)

When Bell began to suffer from ill health, and after the death of his two brothers to tuberculosis, Bell's father moved the family to the more pleasing climate of Brantford, Canada. It was in Brantford, a town about 60 miles west of Niagara Falls, that Bell spent his summers carrying out many of his experiments on the telephone. In the winter he spent his time in Boston, working and researching.

At the age of 25, in October, 1872, Bell opened his own school for the deaf, the "School of Vocal Physiology," on West Newton Street in Boston. He also taught vocal physiology at Boston University. (Beottinger, p.57). While Bell's professional career soared, he continued work on his "harmonic telegraph" attempting to transmit musical notes simultaneously on a single wire. By autumn, 1872, he was creating crude drawings and experiments in connection with this work.

By 1873, however, his course changed and his experiments dealt less with musical sounds and more with the human voice. During the winter of 1873-1874, he worked night and day, non-stop. While lecturing and tutoring pupils by day, he continued his experiments each night. By the summer of 1874, while on vacation in Brantford receiving a much needed rest from his work and research, Bell performed experiments with the membrane of the human ear. His tests showed that a single membrane, rather than the tuned reeds used in his previous experiments, could cause an electric current to vary in intensity to the air waves made by sound, or perhaps even the sound of speech! These experiments were to be the basis for his invention of the telephone. (Brooks, p. 40-41).

Bell returned to Boston to continue tutoring the deaf and experiment with tuning forks, doing all of the mechanical work himself. In Boston's scientific circles, he now was considered an intellectual. As such, he built a lasting relationship with two of his pupils' fathers, Thomas Sanders and Gardiner Hubbard, both prominent men in the Boston business community. As Bell's experiments began to intensify, and his reputation prospered, Sanders and Hubbard took commercial interest in Bell's experiments. Both offered Bell financial backing to apply for a patent, not for the concept of a telephone, but for his "harmonic telegraph." Unfortunately for Bell, neither man backed such a ridiculous idea as that of carrying the human voice along a wire at a distance.

Therefore, Bell continued his work on the harmonic telegraph, and only worked on his telephone on the side. New contacts at the Massachusetts Institute of Technology widened Bell's knowledge of the machinery needed for

GARRET IN THE BUILDING AT 109 COURT STREET, BOSTON

Where Professor Bell and Mr. Watson made their experiments, resulting in the latter hearing the first sound ever transmitted by telephone, on June 2, 1875. The garret at that time was divided into two rooms, the partition being at a point midway between the windows.

(Cut loaned by the State Street Trust Co. of Boston.)

Photograph of the garret of Charles Williams Jr.'s shop building at 109 Court Street in Boston. The site where Professor Bell and Watson made their early experiments together. 4.75" x 6". *Courtesy of Norman M. Mulvey.*

his "telephone." (Paper 29, p. 315). In January, 1875, Bell moved his experiments to the electrical shop of Charles Williams, Jr. at 109 Court Street, in Boston and hired Thomas A. Watson to help him with the mechanical work. Watson, at the age of 18, held a keen interest in spiritualism, giving him the ability to accept Bell's concept of a telephone which, at the time, was still considered an idea from the supernatural. The two began a working relationship which altered technological history. (Boettinger, p. 57).

In June of 1875, Bell and Watson were in the midst of performing another of their numerous experiments, and the two men stood in separate rooms of the attic at the electrical shop testing their harmonic telegraph equipment. Watson would press the transmitting keys causing the

reeds to vibrate at carefully tuned pitches while Bell received the incoming tones. At one point, one of the reeds stuck, so Watson plucked the reed with his finger. Suddenly, Bell screamed from the other room. He had heard the sound of the reed! A screw was too tight, causing the transmitter to produce a steady current which carried the sound of Watson's plucking on the air waves! (Brooks, p. 44-45). After listening to the noise over and over again with Watson, Bell realized that they had actually transmitted and reproduced, electrically, the identical sound made by the plucked reed. He was convinced that other sounds could be reproduced with equal clearness, and perhaps even be made to produce voice waves of sufficient power for practical transmission of speech.

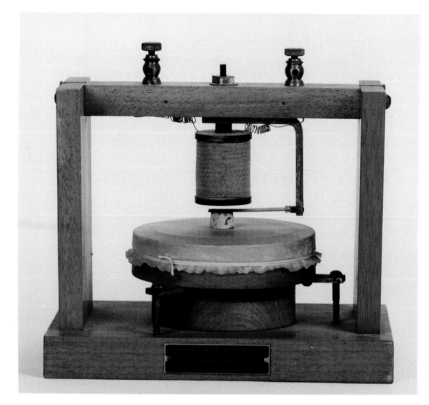

A Bell Company reproduction of Bell's first magneto telephone, the "gallows" telephone from 1875. This is the first patented telephone over which sound was transmitted, but no audible speech was heard. *Courtesy of Private Collection.*

Bell instructed Watson to build a telephone according to Bell's sketches, and to do it as quickly as possible. This first piece of telephone equipment became known as the "gallows" telephone due to the shape of its wooden frame. Watson had it ready for testing by the next evening. This was Bell's first magneto telephone. The instrument consisted of a parchment diaphragm which took in the sounds of the voice and transformed them into electrical pulsations, thereby transmitting speech electrically.

Watson and Bell tested the gallows telephone, and although tone and pitch were audible, no actual speech was transmitted. Watson stated enthusiastically that he could "almost catch a word now and then," but this was a far step from the likes of a working "telephone." (Paper 29, p. 320). The equipment was basically a remodel of Reis's telephone and although essentially a failure, Bell had stepped onto the right path which, after some unexpected detours, led him to invention. (Brooks, p. 45).

Bell realized that he was not the only one performing such investigations in this age bursting with invention and science. At about the same time, Elisha Gray, superintendent of the Western Electric Manufacturing Company and an expert electrician, also had ideas about a "harmonic telegraph." By May of 1874, Gray had constructed a transmitter with eight metal reeds which could vibrate to correspond to the eight notes of a diatonic scale. He worked on a receiver as well which would receive the notes of the transmitter.

Gray applied for a patent on his harmonic multiple telegraph system on June 28, 1875, and received the patent on July 20, 1875. His successful experiments led him to further experiments on transmitting the human voice. He reasoned that, if multiple tones could be sent at the same time over a telegraph wire, then why couldn't the human voice also be sent? (Paper 29, p. 46-47).

Bell, aware of a serious contender with whom he must compete, was led to intensify his experiments. Fearing the security of his research, Bell moved his office to two rooms in a boarding house at 5 Exeter Place, Boston where Bell and Watson would carry out the rest of their experiments.

At the end of 1875, Bell fell ill once again. He and Watson had worked feverishly throughout the year, and Bell was physically, as well as financially, drained. He returned to his father's home in Brantford to regain his health. In

Brantford, he wrote the details for the patent filed on February 14, 1876. In this description, two methods for transmitting sound were discussed. The way in which they were written on the application later caused controversy. Some 600 patent suits were eventually fought and won by Bell and his associates.

Whether it was fate, or just pure chance, Bell's greatest contender for the invention of the telephone, Elisha Gray, filed a caveat for his method of transmitting and reproducing speech on the same day that Bell filed his patent, February 14, 1876. By just a few hours, Bell was able to precede Gray and win rights to the discovery of one of the greatest technological advances of all time. Ironically enough, when Bell made his application for the patent, nothing even close to a "telephone" had been developed. In fact, the patent itself was titled "Improvements in Telegraphy" with no mention of the word "telephone." (Paper 29, p. 320).

Bell's patent came under scrutiny, not just because of his timing, but, more important, for what the patent described. Bell's patent mentions two types of electrical transmitters of "vocal or other sounds." The first consisted of Bell's magneto-induction principle, the type which failed to transmit voices in Bell's experiments on his "gallows" telephone equipment. With this method, a membrane was placed near an electromagnet. The second method used the variable resistance principle, in which the current was charged from a battery. (Paper 29, p. 320).

The variable resistance principle caused most of the controversy. Such a transmitter was described in Gray's caveat, filed just hours after Bell's patent. Bell, who evidently had never experimented with the variable resistance principle, included the principle only in the margin of the page on his application, whereas his magneto transmitter received a full page discussion. In this sense, it appears as though the variable resistance was simply an afterthought, added at the last minute. When asked why it was written in such a way, Bell responded that, "almost at the last moment...I discovered that I had neglected to include...variable resistance." So he added it to his original thoughts.

Another concern was Gray's inclusion of acidic water to transmit sound in a continuous variable resistance. Bell had never experimented with this type of transmitter before filing application, and within a month after application he was testing the method using sulfuric acid such as described in Gray's caveat.

It will always be a mystery, therefore, as to who really provided the basis for the invention, Bell or Gray. But because Bell legally held the patent rights, he continued his work and eventually invented the first telephone.

Bell continued experimenting with his magneto telephone, to no avail. He switched his methods and used the variable resistance transmitter mentioned in both Gray's caveat and his own patent. He replaced the electromagnet in his magneto telephone with a dish of water containing sulfuric acid. By March 10, 1876, Watson had built a transmitter which consisted of a small wire, a diaphragm and diluted sulfuric acid. The vibration of the voice caused the wire to dip into the acid dish and rise and fall in the acid by the variance of the vibrations. The current from the battery flowed through the wire and caused the receiver's membrane to vibrate and produce sound. They set up the equipment for testing, and Watson walked into the other room with the receiver in hand. As he lifted the receiver to his ear, he was amazed to hear Bell's voice at the other end exclaiming those simple, yet now so famous, words, "Mr. Watson, come here, I want you!"

Evidently Bell had spilled some of the acid on his clothing, burning himself. He screamed for Watson to come help him without even realizing that Watson had heard, and the telephone had worked for the first time! So the telephone proved from the very beginning to be an extremely useful instrument. The telephone entered into life, not as an elitist instrument for eloquent terms, but as a device used for a call of help. At the age of 29, Alexander Graham Bell became the lawful inventor of this first telephone. (Boettinger, p. 66).

Chapter Two
The First Telephones

Bell vs. Western Union

"Hoy, Hoy!" as Bell wished to be greeted when answering the telephone, never quite caught the public's attention. (Brooks, p. xi). But the telephone itself was soon to catch on in earnest. The 1876 Centennial Exposition, in Philadelphia, was chosen as the place to unveil the invention. Bell and Watson had a lot of work to do between March 10, and the opening of the exhibit on June 25.

After many abortive attempts using the new variable resistance telephone, Bell reverted to his original magneto telephone. The magneto system had the advantage of using the same piece of equipment both as the transmitter and the receiver, with one slight disadvantage, however; shouting was necessary to be heard. The irony continued, therefore, as Bell moved from a superior system (variable resistance) to an inferior one (magneto) and his magneto equipment would be the first to work at the Exhibition in Philadelphia. (Brooks, p. 50).

By then, Bell was engaged to be married to Mabel Hubbard, a painter, whose father, Gardiner Hubbard, had given Bell financial support for his patent on the harmonic telegraph. Mabel, who had been Bell's student in Boston, became deaf at the age of five. Mabel is brought to mind with the saying, "behind every great man is a great woman." According to some reports, she was the inspiration behind Bell's invention of the telephone. Of course it was Mabel who prodded Bell to go to the Exhibition in Philadelphia as well. (Beottinger, p. 82-83).

In Philadelphia, the air was filled with the excitement of invention and celebration. The Centennial was staged in the midst of a surging momentum of discovery and transition. Exhibits were set up in the arts, and in science and technology and fairs were held at both the social and educational levels. Everyone seemed to be touched by the event, especially Bell and his new equipment.

Bell arrived to Philadelphia's sweltering summer heat in plenty of time to meet some of the judges, including Sir William Thomson, known as Lord Kelvin, one of the most famous physicists of all time. On June 25, Bell would have his chance to exhibit his equipment.

On the appointed day, the judges worked hard and were about to leave without seeing Bell's exhibit. Suddenly, Dom Pedro II, the English-speaking emperor of Brazil, noticed Bell and his equipment in an obscure corner of the exhibit hall. The convivial Dom Pedro II went over to greet Bell whom he had met in Boston at Bell's school for the deaf, and the other judges followed suit. (Beottinger, p. 83).

The lasting impression Bell had left with Dom Pedro II provided the chance he needed to publicize his invention. With Dom Pedro listening on the receiving line, and Lord Kelvin and Elisha Gray, among those who stood to watch, (Gray was at the exhibit displaying his own apparatus), Bell walked into another room in the Hall about a hundred yards away and recited from *Hamlet* through the transmitter. Dom Pedro screamed as he heard the words and exclaimed, "I hear, I hear!" When Lord Kelvin took his turn at the receiver, he heard Bell speaking and excitedly ran to find Bell while screaming, "I must see Mr. Bell!" Upon finding him, Bell was speaking into the transmitter saying, "Do you understand what I say?" upon which Lord Kelvin, between gasps of excited breath, exclaimed that he did. (Beottinger, p. 83-84).

Lord Kelvin later reported that Bell's equipment was, "perhaps the greatest marvel hitherto achieved by the electric telegraph." The Centennial celebration was the cause which launched the telephone more quickly into commercial use. Bell was now more determined than ever, and was supported for the first time by other experts in the field. (Brooks, p. 51-52).

Bell and Watson returned to Boston to devote

A Bell Company reproduction of Bell's liquid transmitter and tuned reed receiver. These were used on the great day of discovery, March 10, 1876 as Bell's voice was heard by Watson over the phone when Bell spilled sulfuric acid. *Courtesy of Private Collection.*

all of their time to the improvement of the telephone. In October, 1876, Bell replaced the steel reed with a steel plate almost as large as the membrane in the receiver. At about the same time, Watson learned of a permanent magnet while reading at the library. It replaced the soft iron core of the electromagnet. (Paper 29, p. 321). The improvements these two changes made were quite evident, and on October 9, 1876 the first two-way long-distance call was set up between Boston and Cambridgeport, Massachusetts. Bell and Watson performed their now routine victory dance, which Bell had learned from the Canadian Indians. (Beottinger, p. 88).

In the autumn of 1876, Gardiner Hubbard offered Bell's patents to the Western Union Telegraph company for $100,000. This seems unbelievable now, but for Bell, $10,000, a 10% share, was unimaginable. Blindly fortunate for the Bell team, the deal was refused by Western Union Telegraph.

Bell continued to improve his telephone, replacing the membrane in the receiver with an all-metal diaphragm. He placed the new structure in a box which resembled a photographer's camera of the 1870s. The phone

is now referred to as Bell's "box" telephone. In January of 1877, Bell was issued a second patent for his latest telephone. This final patent was Bell's last technical contribution to telephony as well as the last protection on his receiver-transmitter instrument and other mechanical features from a barrage of future attacks. (Paper 29, p. 322).

On April 4, 1877, the first permanent outdoor telephone wire was placed between the Williams Shop at 109 Court Street and William's home in Somerville, a three mile distance. On May 1, the first box telephones, or "camera" phones, were rented for commercial use in Massachusetts. As the telephone became commercially available, Bell and Watson were able to afford to eat. However, they were inventors, not businessmen, and they would need help controlling this industry which was on the threshold of expansion and recognition.

Camera phone, c. 1877, the first commercial telephone. The round camera-like opening served as both receiver and transmitter. 7.5" x 8" x 10.5". *Courtesy of Norman M. Mulvey.* $2000-4000.

After Western Union turned down the offer to buy rights to the telephone, Bell and Watson set out on a different course to reap the financial benefits of their research. They began a series of vaudeville-type presentations in New England to demonstrate the telephone. Watson would sing popular songs into the receiver as the audience listened on various telephones hanging about the theatre. One time, as Watson and Bell were practicing for one such show, with Watson in their home/laboratory and Bell in New York, their landlady complained of the noise. Watson covered barrel hoops with blankets, and climbed inside to continue his sweaty, yet wondrous repertoire, with no disturbances from the landlord. Such was the beginning of the telephone booth.

Bell's financial reputation was now stable enough for his long awaited marriage to Mabel Hubbard. The couple married on July 11, 1877. From that time on, Bell left the scientific and managerial aspects of telephony to Watson and Gardiner Hubbard and devoted his time to promoting the telephone in Europe. Watson and Hubbard, meanwhile, formed the Bell Telephone Company with Hubbard as trustee. Charles Williams' shop, where Bell and Watson had begun their experiments and where Watson had been employed, became the manufacturer for the telephones under Watson's directions. The company was on its way, lacking in only one respect. They still did not have a sufficient transmitter. (Beottinger, p. 94, 96).

The telephone did not take such an advanced course in Europe. Large-scale telephony as was known in the United States in the late 19th century, was to await the coming of the turn of the century in Europe. New York state had almost as many telephones in operation by the turn of the century as did all of Europe! Also, the telephone in Europe was run by the government. Rates were often so low that they could not afford to keep up the equipment.

In America, the response to the telephone was explosive despite its primitive equipment. People accepted the idea of communicating across wires as they saw neighboring businesses substitute the telephone for telegraph instruments. Western Union, realizing the mistake of not buying Bell's patents, competed with the Bell company by producing their own phones despite patent infringement. Ironically, this helped rather than hindered the Bell company by making the instrument more reputable. As the public saw an established company such as Western Union investing in telephones, the leasing of telephones escalated, for both Western Union and the Bell Company. On June 30, 1877, 230 telephones were in use. By July, that figure had risen to 750 and by the end of August it was up to 1,300. (Ithiel de Sola Pool).

American businesses suddenly realized that the ''American Dream'' was attainable through the telephone and the newly formed Bell Company was hit by a plethora of competitors all trying to reap the benefits of the technological discovery.

Elisha Gray was upset that Bell was receiving all of the credit. Another man, Professor Amos E. Dolbear of Tufts College, was also trying to fight the Bell patents. He stated that, in 1876, he had made a sketch which improved upon Bell's phone. Both Gray and Dolbear made an agreement with Western Union to challenge Bell's patents, arguing that Gray's caveat, combined with Dolbear's work, constituted a fair challenge to the Bell patents. Western Union was on its way in trying to win back the chance to own the telephone, the chance it had given up just one year earlier.

In December of 1877, Western Union set up the American Speaking Telephone Company to conduct its telephone business controlled by another of its subsidiaries, the Gold and Stock Telegraph Company. American Speaking would become Bell's first and biggest competitor from 1877-1879. Holding the legal rights to Gray's and Dolbear's claims, and with G.M. Phelps' crown receiver, they sought to strengthen their hold in the telephone market. Ultimately, they found the Bell Company's weakest point, the transmitter, and hit them where it hurt. Although Bell had improved his receiver, his magneto transmitter was not adequate for commercial use. Shouting was necessary and was only audible at short distances. If the Bell Company was to survive, they needed to produce a stronger transmitter, and fortunately for Western Union, they had just the man to do the job.

Western Union had hired Thomas Alva Edison to its forces in 1876. He had researched the telephone that year and by the middle of 1877 had developed a carbon button transmitter which proved much more efficient than Bell's magneto transmitter, eliminating the need to shout in order to carry out a conversation. Edison's patent precluded anyone from using carbon in a transmitter, or so Edison thought.

The Bell Company did not simply kick back

and watch as Edison did research for Western Union. Instead, Watson hired the company's first general manager, Theodore N. Vail, a man who proved to be a great inside strength to the business. Vail became the leading force as the company took a stance against the ensuing patent suits. With the help of Emile Berliner, a German immigrant, who had spent some time experimenting with electricity, the Bell company also developed its own improved transmitter. Berliner's transmitter was similar to that of Reis's, but of a greater durability. Although the Berliner transmitter was not used much commercially, it helped the Bell Company to impede Western Union and Edison's transmitter patent.

Edison recognized the limitation of the carbon transmitter to function without a variable resistance, so he devised a means of variance in a patent issued April 30, 1878. Emile Berliner devised and filed a patent covering the use of an induction coil with a transmitter in January of the same year. This action allowed the Bell Company to file an interference against Edison's patent. The Berliner interference also protected the Bell company from an injunction by Edison against its use of the Blake carbon transmitter and that immunity was a major factor in the final settlement with Western Union on November 10, 1879, which gave the Bell Company Edison's telephone rights. Edison protested Blake's use of carbon in a transmitter and interference against his transmitter.

Edison won his patent after long drawn out proceedings in 1892, but by then it was too late. Perhaps this was the reason that all Blake transmitters during this period carried all the names and patent dates as well as the patent numbers of all the people involved -Bell, Berliner, Edison, and Blake. (Paper 29, p. 328-329) (ATCA Fact Sheet).

Common battery wall phone. All of the names were written on the transmitter at the time of the suits over Edison's carbon transmitter. *Courtesy of James P. Barr.*

Watson's "thumper" telephone, c. 1877. Before a bell existed on the telephone, one had to alert the person at the other end by tapping on the mouthpiece with a pencil. Watson added the small knob below the mouthpiece to act as a hammer signal or "thumper." *Photo courtesy of Private Collection.*

"Iron box," c. 1876. The receiver used by Dom Pedro was similar to this cylindrical iron box with a magneto core and lid diaphragm. Bell used such a receiver and a membrane telephone to talk from Brantford to Paris, Ontario, about an 8 mile distance, the first time articulate speech was transmitted and received at so many miles. *Photo courtesy of Private Collection.*

"Butterstamp" telephone, c. 1877. Used as both transmitter & receiver, this is the first commercial hand telephone. This particular telephone belonged to Queen Victoria with a British patent no. 67. *Courtesy of Private Collection.* $1500-2500.

Coffins

Edison transmitter, c. 1878. *Courtesy of Bill Elsasser.*

Coffin telephone with magneto transmitter, c. 1878-1879. Made by Charles Williams, Jr. for the National Bell Telephone Co. Walnut, 12'' x 23'' x 5.5''. *Courtesy of Norman M. Mulvey.*

Coffin telephone c. 1878, mounted on display. Made by Charles Williams, Jr. for the Bell Telephone Company. Walnut, 11'' x 11'' x 6''. *Courtesy of Norman M. Mulvey.*

Bell coffin telephone with magneto transmitter mounted on board with instruction sheet, c. 1877-1879. Mahogany, 48'' x 13.5''. *Courtesy of Norman M. Mulvey.*

* Page 17,18&19: All $1500-5000 ea.

Coffin telephone with two original wooden Bell receivers!, c. 1877-1878. Made by Charles Williams, Jr. for the Bell Telephone Co. Walnut, 15'' x 5'' x 5''. *Courtesy of Norman M. Mulvey.*

Coffin telephone, c. 1878. A two line telephone made by Charles Williams, Jr. for the National Bell Telephone Co. Walnut, 12.5'' x 7'' x 5''. *Courtesy of Norman M. Mulvey.*

Coffin telephone, c. 1878. Made by Charles Williams, Jr. for the National Bell Telephone Co. Berliner transmitter and Roosevelt automatic switch hook. Mahogany, 13'' x 5.25'' x 5.5''. *Courtesy of Norman M. Mulvey.*

Last model of the coffin telephone, c. 1879. Equipped with a Blake transmitter and Roosevelt automatic switch hook. Made by Charles Williams, Jr. for the National Bell Telephone Co. Mahogany, 13'' x 5'' x 6.75''. *Courtesy of Norman M. Mulvey.*

Coffin telephone. Made for the National Bell Telephone Co. *Courtesy of Norman M. Mulvey.*

Coffin telephone with two wooden bells, 1877-1878. Made by Charles Williams, Jr. for the Bell Telephone Co. Oak, 13.5'' x 5.75'' x 5''. *Courtesy of Norman M. Mulvey.*

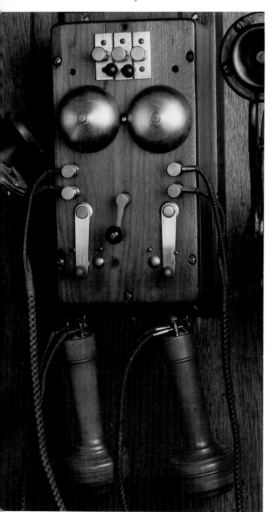

Bell coffin telephone made by Davis & Watts, of Baltimore, c. 1878. Mahogany. *Courtesy of Norman M. Mulvey.*

Watts telephone, c. 1878. Made for the National Bell Telephone Co. Walnut, 14'' x 6.25'' x 4.5''. *Courtesy of Norman M. Mulvey.*

Coffin telephone, c. 1878. Made by Davis & Watts, Baltimore, Maryland for the National Bell Telephone Co. Equipped with knobs for wires and a mechanical switch hook. *Courtesy of Norman M. Mulvey.*

Coffin telephones, c. 1879. Made by the Post & Co. for the National Bell Telephone Co. One of burl wood, 12'' x 6.5'' x 5.25''. *Courtesy of Norman M. Mulvey.*

Coffin telephone, c. 1879. Made by the Post & Co. for the National Bell Telephone Co. 12'' x 6'' x 5''. *Courtesy of Private Collection.*

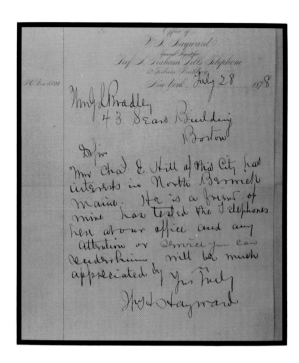

Telephone with Edison transmitter and G.M. Phelps receiver, c. 1878. Walnut, 8.75'' x 8'' x 7.5''. *Courtesy of Bill Elsasser.* $1500-3000.

Sales brochure for the Bell Telephone Company dated July 28, 1878.

Western Union telephone, c. 1879. Cherry wood, 7.75'' x 6'' x 4.5''. *Courtesy of Norman M. Mulvey.* $7500-8500.

American Speaking Telephone Company telephone with G.M. Phelps receivers, crank and instructions on front. Patented August 19, 1879, manufactured by Western Electric. Walnut, 1.5'' x 9.5'' x 4.75''. *Courtesy of Norman M. Mulvey.*

American Speaking Telephone Company telephone with an Edison transmitter. First model of this type, c. 1879. 8.75'' x 8'' x 7.5''. *Courtesy of Norman M. Mulvey.*

Edison carbon transmitter, c. 1878. Wood and brass, embossed, 4.5'' x 4.5'' x 2''. *Courtesy of Norman M. Mulvey.*

IMPERATIVE RULES.

1.—Keep the hand Telephone in its place till the calls are answered.

2.—Call and answer by pressing the button *C,* and turning the crank.

3.—When the call is answered, take down the Telephones, and you are ready for conversation.

4.—To **open** the line for use, turn the switch to the right. **Close** it during thunder-storms, and when not required for use, by turning the switch to the left.

Pair of G.M. Phelps receivers, c. 1878. Made for the American Speaking Telephone Co. 5.5'' x 2.75''. *Courtesy of Norman M. Mulvey.*

Close-up of instructions on telephone at left.

Gold and Stock Exchange telephone, with G.M. Phelps receivers, c. 1879-1880. Made by Western Electric. Walnut, 8.25'' x 9.5'' x 4.25''. *Courtesy of Norman M. Mulvey.*

Switchboards

Four line switchboard, c. 1882. Western Electric Manufacturing Co. Burl wood 17" x 9" x 5.5". *Courtesy of Norman M. Mulvey.* $850-1000.

Two parts of an incomplete Charles Williams, Jr. telephone mounted on a board. Blake transmitter. 15.5" x 8.5" x 5". *Courtesy of Norman M. Mulvey.* $1000-1500.

Single line switchboard, c. 1880. Maker unknown. Walnut, 16.5" x 15.25" x 2.75". *Courtesy of Norman M. Mulvey.* $250-550.

Holcomb string telephone with wooden receivers, c. 1880. This phone was used between a house and an ice skating rink in Maine. J.R. Holcomb Co. Walnut, 9.25" x 8" x 7.75". *Courtesy of Norman M. Mulvey.* $1000-1300.

6-line wall switchboard, pat. Nov. 24, 1885. Made by Emmner Imperial Telephone, Washington, DC. Carbon rod transmitter. Walnut, 33" x 17" x 13". *Courtesy of Norman M. Mulvey.* $2000-4000.

Holcomb string telephone patented April 26, 1881. Made by J.R. Holcomb Co. Walnut. *Courtesy of Norman M. Mulvey.* $350-500.

DeVeau switchboard, c. 1894. DeVeau feared a law suit for patent infringement by the Bell company and therefore built this switchboard based on a direct current system instead of the contemporary alternating current system. This made it a much more primitive system for its day. 15.5" x 5.5". *Courtesy of George Howard.* $2000-3000.

String Phones

Lover's telephone. The advertisements for this type of phone had a man at one end and a women at the other speaking on the phone, thereby giving it the name, the lover's telephone. J.R. Holcomb. Left: 4'' x 3.75''. Right: 3.5'' x 2.5''. *Courtesy of Norman M. Mulvey.*

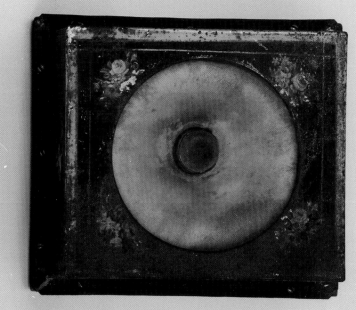

Smaller Watts string telephone with decals. Tin and animal skin, 8.25'' x 6'' x 3.5''. *Courtesy of Norman M. Mulvey.*

Clacker telephone, c. 1880s. Crank for clacker to make noise. J.R. Holcomb. Walnut, 3.75'' x 3.75'' x 3''. *Courtesy of Norman M. Mulvey.*

Anti-Bell string telephone, c. 1880s. Made in Boston. 8.5'' x 8.5'' x 3.75''. *Courtesy of Norman M. Mulvey.*

* Page 24-28: All $100-500 ea.

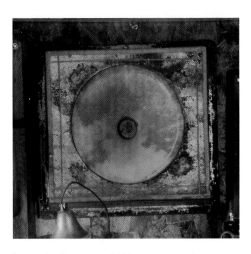

Watts string telephone, c. 1880. Watts Telephone Co. Tin, 10.75'' x 10.75'' x 4.75''. *Courtesy of Norman M. Mulvey.*

String telephone with call bell, c. 1880s. Niles Chair Co., Niles, Michigan. Equipped with double funnel to make it work better. Walnut, 9'' x 8.5'' x 10.5''. *Courtesy of Norman M. Mulvey.*

String telephone, c. 1880s, Ciro, New York. Maple, 5'' x 5'' x 2''. *Courtesy of Norman M. Mulvey.*

String telephone with magneto call bells, c. 1880s. Shaver Multiplex Telephone, New York. Cherry wood, 22'' x 9'' x 6.5''. *Courtesy of Norman M. Mulvey.*

Ornate string telephone which swivels on its hanging stand, c. 1880s. Hulls Jumbo Telephone Co., Boston. Walnut, 9.75'' x 12'' x 9''. *Courtesy of Norman M. Mulvey.*

String telephone with rare wooden receiver, c. 1880s. National Telephone Company, Boston. Walnut, 5'' x 5''. *Courtesy of Norman M. Mulvey.*

Unique string phone in original box with original spool of wire, c. 1880s. Bliss Telephone. Box: 2.75'' x 8.25'' x 3.75''. String phones: 3'' diameter. *Courtesy of Norman M. Mulvey.*

String telephone, c. 1880. Larkin Telephone Company, Westfield, Massachusetts. Cast iron, 4.5'' x 4.5'' x 7''. *Courtesy of Norman M. Mulvey.*

String telephones, c. 1880. Daisy Telephone Co., Akron, Ohio. Walnut, 6'' x 3.75'' x 2''. *Courtesy of Norman M. Mulvey.*

String telephone with call bell and wire which was pulled to ring the bell. Made by Viaduct Electric. Walnut, 14.5'' x 8.25'' x 5''. *Courtesy of Norman M. Mulvey.*

String telephone patented Sept. 10, 1889 by the Elliot Telephone Company, Mooresville, Indiana. Walnut, 6'' x 6'' x 3.75''. *Courtesy of Norman M. Mulvey.*

Harmonic telephone, c. 1880s. The carbon disk inside works on the principal of harmonics. Maker unknown. Oak, 5.75'' x 5.75'' x 1.5''. *Courtesy of Norman M. Mulvey.*

String telephone with call bell, c. 1889. Mechanical Telephone Company. Tin and papier maché, 8'' x 8'' x 2.5''. *Courtesy of Norman M. Mulvey.*

String telephone, c. 1880s. Maker unknown. Rosewood, 15.75'' x 5'' x 4.5''. *Courtesy of Norman M. Mulvey.*

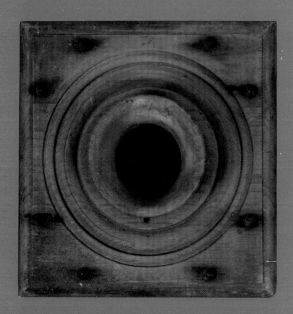

String telephone with call bell on top, c. 1880s. Maker unknown. Walnut, 9" x 8". *Courtesy of Norman M. Mulvey.*

String telephone, c. 1880s. Probably homemade. Pine, 5.75" x 6.25" x 6". *Courtesy of Norman M. Mulvey.*

String telephone, c. 1880s. Maker unknown. Walnut, 12" x 7.75" x 5.25". *Courtesy Norman M. Mulvey.*

String telephone which makes duck noise when blown into, patented Aug. 2, 1881. Maker unknown. Walnut, 8" x 7" x 3.75". *Courtesy of Norman M. Mulvey.*

Tube telephone, c. 1880. Double string phone, yell into the tube. One side is marked "Hospital," the other "Door." 10.5" x 11.75" x 3.25". *Courtesy of Norman M. Mulvey.*

Vanities

The vanity desk set was an elaborate telephone desk used generally in hotels and in the homes of the wealthy. The vanity consisted of a telephone built into a beautiful table so that the person using the phone could sit while they spoke and have the use of a desk.

Non folding vanity, c. 1890s. Western Electric long distance telephone made for Southern New England Telephone, American Bell. Oak, 41" x 27" x 21". *Courtesy of Norm M. Mulvey.*

* Page 29&30: All $2200-5500 ea.

Vanity desk set, c. 1890s. Western Electric. *Courtesy of Private Collection.*

Folding vanity, c. 1892. Western Electric long distance telephone made for Southern New England Telephone. Oak, 41'' x 25'' x 21''. *Courtesy of Norman M. Mulvey.*

Fancy sunburst vanity, c. 1898. Western Electric. Equipped with sound proof mouthpiece, Talk Low Co. of New York. American Bell Telephone Co. Oak, 51'' x 24'' x 17''. *Courtesy of Norman M. Mulvey.*

Very early model Western Electric vanity desk set, c. 1884. Unusual transmitter (replica) first patented March 7, 1876. *Courtesy of Private Collection.*

Chapter Three
Wooden Wall Telephones

Transmitters

The advent of Edison's carbon transmitter caused new inventors to come onto the scene to develop improvements of their own.

The Blake Transmitter

Francis Blake Jr. designed a carbon button transmitter early in 1878. Known as the Blake transmitter, it consisted of a platinum bead, a diaphragm and a carbon block. The diaphragm pushed the bead, which was attached, against the carbon block. The transmitter was sold to the Bell Company. At first it did not work altogether well, so Berliner improved it by substituting a harder carbon block. This newer version of the Blake transmitter proved to be better than Edison's carbon transmitter. The Blake instrument was patented in England on January 20, 1879 and in the United States on November 29, 1881. The Blake transmitter was soon to be standard equipment in Bell telephones and was also used extensively in equipment abroad. (TCI Newsletter Vol. 4, No. 1).

The Hunnings Transmitter

Henry Hunnings, an English clergyman, replaced the carbon in Blake's transmitter with granules of coke which were placed between the diaphragm and a metal support. This was the first carbon grain transmitter. The system was widely used because it resulted in a clearer, louder transmission. The disadvantage of the granular system was that the carbon granules would pack with constant use, resulting in a loss of sensitivity.

Pencil Transmitter

This type of carbon transmitter was more prominent in European telephones and were made by various manufacturers. It is so called because two or more carbon pencils sit in carbon sockets arranged in varying order around the wooden diaphragm. The vibrations strike the diaphragm and spread in various directions.

Blake transmitter, c. 1880s. Shown open and closed. 6.5" x 5.5" x 3.25". *Courtesy of Norman M. Mulvey.* $100-300.

Due to internal conflicts, Western Union weakened in its fight against the infant, yet strong-willed Bell Telephone Company. William H. Vanderbilt, president of Western Union, was experiencing an attempt at takeover by the well-known financier Jay Gould. In 1878, the American Bell Telephone Company, as it was then called, filed suit against one of Western Union's subsidiaries. By 1879, Western Union had fallen weak under the works of Jay Gould. Outside of court Western Union decided to give over all the patents and facilities to Bell, including the Edison transmitter and Gray and Dolbear patents, as well as a network of 56,000 telephones in 55 cities. In exchange, Western Union would receive 20 percent of Bell's royalties for the next 17 years, when the Bell patents would expire. This settlement gave the National Bell Telephone company a legal monopoly over the telephone business that would last until 1893 and 1894, with the expiration of Bell's patents.

Until the spring of 1879, Charles Williams, Jr.'s shop was the sole manufacturer for the Bell Company. It was very difficult for the small shop to supply the great and ever-increasing demand for telephones. That spring, other small shops were licensed to manufacture phones and equipment, including Ezra T. Gilliland of Indianapolis.

Vail, as general manager, kept trying to strengthen the company to prepare for the expiration of Bell's patents. In 1881, under his direction, the American Bell Company bought the controlling interest in Western Electric from Western Union. From 1878-1879, the Western Electric Manufacturing Company of Chicago, formerly Gray & Barton, (Elisha Gray & Enos Barton), had supplied Western Union's telephone equipment. So the American Bell Telephone Company now owned telephone service as well as manufacturing. Western Electric bought both William's and Gilliland's licenses, so in 1882, Western Electric became the sole supplier of Bell equipment and became a part of the Bell system, with Enos Barton directing Western Electric for 40 years.

The 1880s saw an expansion of telephone use. In 1878, Thomas A. Watson devised the polarized ringer, a signaling device to signal between stations and calling the operator. In 1879, H.L. Roosevelt patented the automatic switch hook which notified the operator when a phone was in use. The Bell Company held patents on the automatic switch hook until 1891. Until that point, competitors developed the paddle hook to

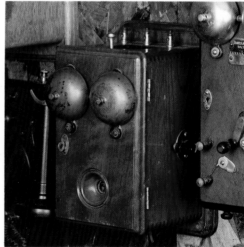

Magneto telephone, c. 1880. Made by Charles Williams, Jr. for the Bell Telephone Company. Open and closed view. Walnut. *Courtesy of Norman M. Mulvey.* $550-750.

circumvent the Bell patent.

The Bell company soon entered an age of patent suits that continued for the next twenty-five years with over six hundred law suits. Theodore Vail as the general manager, led the company through a great many trials and strengthened it as he did so.

Exchanges were beginning to go into service in 1878. The first real exchange went into service in New Haven, Connecticut of that year connecting 21 subscribers. The subscribers did not have numbers yet, so they were called by their name. The first operators for these exchanges were boys. Teen-age boys, however, were soon found to be too impatient with no attention span. In September, 1878, Emma M. Nutt was hired as the first woman operator and the male operator was not seen again until the 1960s.

Four manufacturers at the turn of the century, c. 1895-1905, are considered to be leaders in terms of quality and quantity in the telephone industry. Western Electric, Stromberg-Carlson, Kellog and American Electric (later to be Automatic Electric) all produced mass quantities of quality telephones.

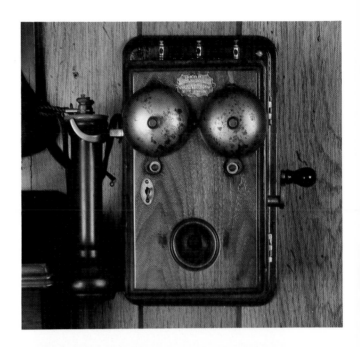

Magneto telephone, c. 1880s. Made by Charles Williams, Jr. for the American Bell Telephone Company. View of inside and close-up of company tag. Walnut, 10" x 6.25" x 8.25. _Courtesy of Norman M. Mulvey._ $550-750.

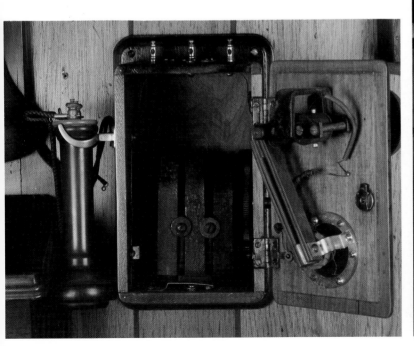

Western Electric telephone, c. 1880. Walnut, 13" x 6" x 4.5". *Courtesy of Norman M. Mulvey.* $600-800.

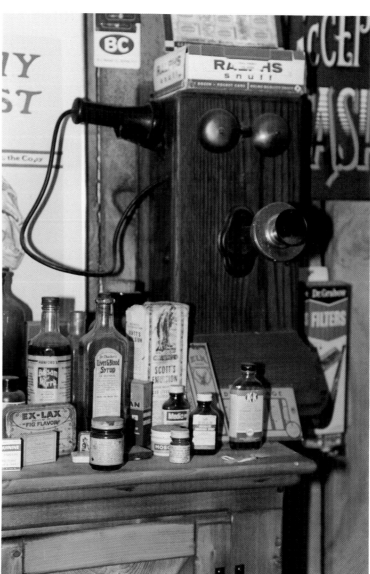

The telephone in its origins was often seen in the drugstore or the rural country store. *Courtesy of Todd General Store.*

Compact telephone, patented 1880. Uses a Blake transmitter not shown. Made by Charles Williams, Jr. for the American Bell Telephone. Walnut, 13'' x 6'' x 7''. *Courtesy of Norman M. Mulvey.* $750-850.

Blake three box telephone, c. 1881. The Blake transmitter is the middle box, a carbon transmitter that improved upon Edison's carbon transmitter. Made by Davis & Watts. Close up of original cherub decoration. Walnut, 28" x 7" x 8.25". _Courtesy of Norman M. Mulvey._ $1800-2000.

Three box telephone with a Blake transmitter, c. 1883. Viaduct Manufacturing Co., (formerly Davis & Watts.) made for the American Bell Telephone Co. Walnut, 27.25" x 8.5" x 8". _Courtesy of Norman M. Mulvey._ $1800-2500.

Wooden box desk stand Aug. 1894. Made by Western Electric for the American Bell Telephone Company. 12" x 5.75". _Courtesy of Norman M. Mulvey._ $650-850.

* Page 36,37&38: All $1500-2000 ea.

Compact three box telephone, c. 1881. Made by Charles Williams, Jr. for the American Bell Telephone Company. Walnut 33'' x 7.5'' x 6.25''. *Courtesy of Norman M. Mulvey.*

Three box telephone with mechanical switch hook, c. 1881. Made by Charles Williams, Jr. for American Bell. 31.5'' x 8.75'' x 6.5''. Inside view to show switch hook. *Courtesy of Norman M. Mulvey.*

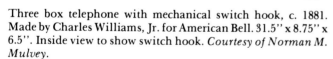

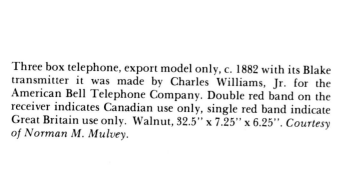

Three box telephone, export model only, c. 1882 with its Blake transmitter it was made by Charles Williams, Jr. for the American Bell Telephone Company. Double red band on the receiver indicates Canadian use only, single red band indicate Great Britain use only. Walnut, 32.5'' x 7.25'' x 6.25''. *Courtesy of Norman M. Mulvey.*

Blake three box telephone with the crank in front, c. 1882. Manufactured by Western Electric for the American Bell Telephone Company. Walnut and cast iron, 31'' x 8'' x 6.5''. *Courtesy of Norman M. Mulvey.*

Three box telephone, c. 1882. Made by Standard Electric for the American Bell Telephone Company. Walnut, 32.25'' x 7.25'' x 6''. *Courtesy of Norman M. Mulvey.*

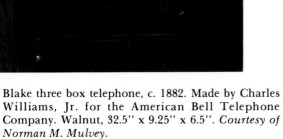

Blake three box telephone, c. 1882. Made by Charles Williams, Jr. for the American Bell Telephone Company. Walnut, 32.5'' x 9.25'' x 6.5''. *Courtesy of Norman M. Mulvey.*

Western Electric newspaper advertisement, March 22, 1883. Copyright by Delano & Co., New York. Electrical Review. Vol. 2 No. 3. Article on an experiment with the telephone. 15.5'' x 22.5''. *Courtesy of Norman M. Mulvey.*

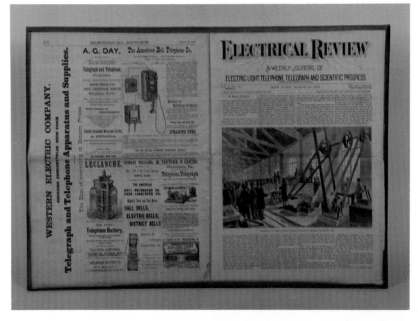

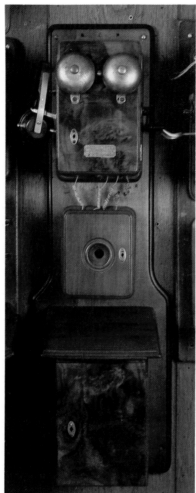

Blake three box telephone, patented 1883. Canadian Bell Telephone Co. of Canada, Montreal, Canada. Walnut. *Courtesy of Art Hyde.*

Blake three box telephone. Manufactured by Gilliland Electric, Indianapolis, for the American Bell Telephone Company. Burl wood, 30'' x 10'' x 6.75''. Headset c. 1884, by Richardson, probably for the railroad. *Courtesy of Norman M. Mulvey.*

Blake three box with whisper phone attached, c. 1883. No springs on hinges. Manufactured by Western Electric. Walnut, 31.5'' x 9'' x 7.75''. Whisper phone made in Cleveland, Ohio, patented Dec. 6, 1887. The whisper phone was an added attachment for privacy so one could speak in a whisper, generally used on three box phones. *Courtesy of Norman M. Mulvey.*

American Bell two box telephone, c. 1884. Made by Gilliland Electric. Burlwood and walnut, 16.75'' x 7.25'' x 6.75''. *Courtesy of Norman M. Mulvey.*

Blake three box, c. 1886. Made by Western Electric for the American Bell Telephone Co. Springs on hinges patented in 1886. Walnut, 31.5" x 8" x 8". *Courtesy of Norman M. Mulvey.* $1500-2000.

Western Electric magneto telephone c. 1891. Equipped with a watchcase receiver. Walnut, 9" x 5.5" x 6.25". *Courtesy of Norman M. Mulvey.* $500-800.

Compact three box telephone, c. 1888. Made by Western Electric for the New England Telephone Co. Equipped with mounted speaking part & unique Blake transmitter. Walnut, 23.5" x 7" x 9.75". *Courtesy of Norman M. Mulvey.* $1800-2000.

Two box telephone, c. 1880s. Unknown maker, probably for Bell Telephone. Oak, 20" x 7" x 7". *Courtesy of Norman M. Mulvey.* $1500-2000.

Gilliland Electric telephone, c. 1894. Equipped with the unique spoon receiver shown. Chestnut, 24.5" x 7.75" x 8.25". *Courtesy of Norman M. Mulvey.* $800-1000.

Western Electric long distance telephone, c. 1895. Made for the American Bell Telephone Co., holds three batteries. Walnut, 34" x 14.5" x 13". *Courtesy of Norman M. Mulvey.* $1000-1250.

American Bell two box telephone, c. 1895.l Made by Western Electric. Equipped with No. 5 arm (early model) with coil behind arm. The No., 5 arm was made from 1891-1895 and was replaced with a smaller arm with the coil in the top box. *Courtesy of Norman M. Mulvey.* $750-1000.

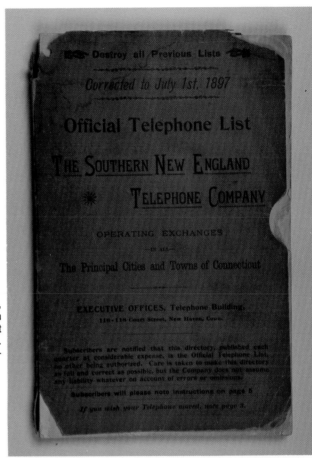

Western Electric long distance telephone, c. 1895. Holds three large batteries. When the batteries sit on top of one another, it is called a tandem telephone. Oak, 6'. *Courtesy of Norman M. Mulvey.* $1000-1250.

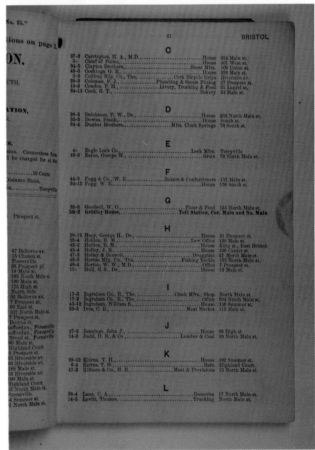

Early telephone book in Connecticut, July, 1897. *Courtesy of Norman M. Mulvey.*

American Bell "fiddleback," c. 1898. Made by Western Electric. Local battery, walnut. *Courtesy of Norman M. Mulvey.* $350-550.

American Bell two box telephone, c. 1902. Made by Western Electric. Walnut, 30.25" x 7.5" x 9". *Courtesy of Norman M. Mulvey.* $350-550.

Western Electric 301, c. 1902. Walnut, 18" x 9.5" x 8". *Courtesy of Norman M. Mulvey.* $250-350.

Western Electric 1317 single box, c. 1902. Picture frame cathedral top. Oak, 22.75" x 9" x 15". *Courtesy of Norman M. Mulvey.* $250-350.

Western Electric common battery wall phone, c. 1905. Oak, 20" x 8" x 13.25". *Courtesy of Norman M. Mulvey.* $250-350.

Western Electric 293A, c. 1910. Oak, 9" x 6.25" x 3.75". *Courtesy of Norman M. Mulvey.* $100-300.

"Donut" phone, also known as a "pancake" phone, c. 1905. Western Electric. *Courtesy of Larry Garnatz.* $175-200.

Western Electric 305G, c. 1910. Walnut, 10.5" x 7" x 5.75". *Courtesy of Norman M. Mulvey.* $100-200.

Western Electric train set wall phone. 21.5" x 8" x 6". *Courtesy of Charles E. Stanley.* $150-225.

Western Electric wall phone with watchcase receiver. 10'' x 6.5'' x 5.5''. *Courtesy of Private Collection.*

Western Electric emergency telephone to call the fire department, 21" x 7.5". *Courtesy of Private Collection. $125-225.*

Western Electric 10-003 hanging handset. One of the first American handsets. 9.25" x 6.5" x 5.25". *Courtesy of Norman M. Mulvey.$100-200.*

Blake common battery, Bell Co. of Canada, Montreal. **OST** (outside terminal) double red band receiver. 8" x 6". *Courtesy of Private Collection. $150-350.*

Ship-to-ship telephone with words imprinted as follows, "For use by his majesty's Admiral only," c. 1880s. Blake transmitter, a National Telephone Company limited. Teak wood, 7.75" x 23" x 8.5". *Courtesy of Norman M. Mulvey.* *$1500-2000.*

Magneto telephone with a clockwork mechanism, c. 1880. Maker unknown. Open view above. Oak, 11" x 6.75" x 3.5". *Courtesy of Norman M. Mulvey.* *$1000-1500.*

Rare telephone, patented in 1880. Made by John Irwin, Morton, Pennsylvania. 6'' x 5.25'' x 4.5''. *Courtesy of Norman M. Mulvey.*

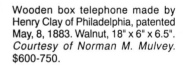

Two box telephone with magneto transmitter, c. 1890s. Maker unknown. Walnut, 22.5" x 7.75" x 6.5". Close-up of inside of transmitter. *Courtesy of Norman M. Mulvey.* $300-400.

Wooden box telephone made by Henry Clay of Philadelphia, patented May, 8, 1883. Walnut, 18" x 6" x 6.5". *Courtesy of Norman M. Mulvey.* $600-750.

Harmonic transmitter, c. 1880s. Never worked and extremely rare. 12" x 13.25" x 4". *Courtesy of Norman M. Mulvey.* $300-400.

Magneto telephone (magnet in back), c. 1880s. Probably homemade. Cherry wood, 12.5" x 3.5" x 2". *Courtesy of Norman M. Mulvey.* $150-250.

Cherry wood telephone patented March 25, 1884. National Improved Telephone Co. 16.25" x 5.5" x 6.25". *Courtesy of Norman M. Mulvey.* $550-750.

Three box telephone with wooden ear piece, c. 1884. Overland Telephone Company. Walnut, 32.5" x 7.5" x 7". *Courtesy of Norman M. Mulvey.* $800-1500.

Three box telephone with carbon pencil transmitter and Molecular receiver, patented Jan. 29, 1884. Made by Viaduct Telephone Co. for the Molecular Telephone Co., NY. Mahogany, 29.5" x 10" x 7.25". *Courtesy of Norman M. Mulvey.* $800-1500.

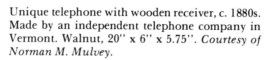

Unique telephone with wooden receiver, c. 1880s. Made by an independent telephone company in Vermont. Walnut, 20'' x 6'' x 5.75''. *Courtesy of Norman M. Mulvey.*

Unique telephone marked R.V.B., Brooklyn, New York. Oak, 16" x 5" x 6". *Courtesy of Norman M. Mulvey.* $500-600.

Dean Oriator with cameo top, c. 1900. Walnut, 29'' x 9'' x 8''. *Courtesy of Norman M. Mulvey.*

Wooden wall phone made by an unknown independent telephone company, c. 1890s. Unique carbon ball transmitter. Walnut, 23" x 7" x 7". *Courtesy of Norman M. Mulvey.* $600-800.

Common battery "fiddleback," c. 1902. Made by the Whitman & Couch Mfg. Co., Boston. Walnut, 20.5" x 7.5" x 7.25". *Courtesy of Norman M. Mulvey.* $350-550.

Top loader singled box, c. 1900. North Electric Co. Oak, 24" x 7" x 13". *Courtesy of Norman M. Mulvey.* $400-500.

Chicago Telephone Supply Company three box, c. 1894. Equipped with long pole receiver and rattlesnake cord. Manual switch for receiver. Golden oak. *Courtesy of Paul McFadden.* $400-600.

Tandem wall phone, c. 1900-1905. Chicago Telephone Supply Co. Glass front. 42" x 9.5" x 7". *Courtesy of Barry Erlandson.* $350-500.

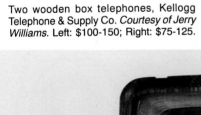

Two wooden box telephones, Kellogg Telephone & Supply Co. *Courtesy of Jerry Williams.* Left: $100-150; Right: $75-125.

Very ornate "fiddleback" telephone, c. 1902. Made by the Swedish-American Telephone Co. Triplet assembly—the receiver hangs up on the arm or speaking device. Oak, 14.75" x 10.25" x 6.75". *Courtesy of Norman M. Mulvey.* $800-1000.

Swedish-American "fiddleback", c. 1906. Trade name was "American Beauty." Triplet assembly. Oak, 27.5" x 11.5" x 13". *Courtesy of Norman M. Mulvey.* $550-750.

Compact two box telephone, c. 1902. Connecticut Telephone & Electric Co. Oak. *Courtesy of Norman M. Mulvey.* $400-600.

The Connecticut Telephone & Electric Co. telephone equipped with watchcase receiver. 5" x 5". *Courtesy of Private Collection.* $35-45.

Stromberg-Carlson experimental model, c. 1894. Equipped with egg-shaped bells, magneto. 12" x 5.25". *Courtesy of Private Collection.* $150-300.

"McKinley" phone, c. 1900. Figure eight telephone made by Stromberg-Carlson. 18" x 9". *Courtesy of Private Collection.* $1000-1500.

Phoenix Electric Telephone Co. telephone, c. 1894. Equipped with Edison transmitter. Oak, 15" x 13.5" x 12". *Courtesy of Norman M. Mulvey.* $1000-1200.

Single box with beading and fancy top made by Williams Electric. Oak, 18.5" x 7.25" x 14.75". *Courtesy of Norman M. Mulvey.* $300-400.

Single box tandem, c. 1902, unique in shape and size. Williams Electric. Oak, 28" x 9" x 13.5". *Courtesy of Norman M. Mulvey.* $400-600.

Long distance telephone, c. 1898. Phoenix Electric Telephone Co. Oak, 32" x 19" x 12". *Courtesy of Norman M. Mulvey.* $400-600.

Blake three box telephone, c. 1891. Made by the Manhattan Electric Co. equipped with paddle switch hook. Walnut, 27.5" x 7.25" x 7.75". *Courtesy of Norman M. Mulvey.* $800-1000.

Close up of the paddle switch hook on the telephone to the left. *Courtesy of Norman M. Mulvey.*

Two box paddle phone, c. 1895. Viaduct Telephone Co. White milk glass mouthpiece is unique. Walnut, 24" x 7" x 8.5". _Courtesy of Norman M. Mulvey._ $600-800.

Magneto telephone, c. 1889. Manhattan Electric. Walnut, 33" x 8" x 7.5". _Courtesy of Norman M. Mulvey._ $500-600.

Two box telephone, c. 1890s. Made by Viaduct Telephone Co. Magneto transmitter & mechanical switch hook. Walnut, 21.5" x 7.5" x 9.75". _Courtesy of Norman M. Mulvey._ $700-1600.

Left: two box paddle phone, c. 1898. Viaduct Telephone Co. Oak, 29.25" x 12.75" x 12". Right: Blake three box, c. 1896, Manhattan Electric Co. Oak, 27" x 8.75" x 7.5". _Courtesy of Norman M. Mulvey._ $500-1200 ea.

Three box paddle phone, c. 1894. Made by American Electric. Bell had the patent on the switch hook so other companies got around it with the paddle. The paddle was held down to talk and to hear. 30.5" x 7.5" x 6.25". *Courtesy of Norman M. Mulvey.* $800-1000.

American Electric "swing away," c. 1902. By 1910 American Electric did away with ornate telephone bodies such as with this picture-frame cathedral top. Walnut, 25" x 9.5" x 15". *Courtesy Norman M. Mulvey.* $300-500.

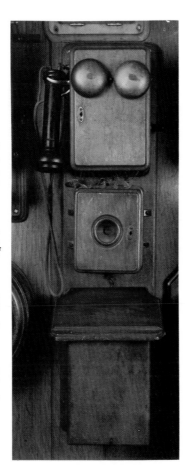

Opposite page center:
Three box telephone, c. 1895. Made by the Keystone Telephone Co.l, Pittsburgh, equipped with Hunnings transmitter. Oak, 31" x 8" x 8". *Courtesy of Norman M. Mulvey.* $500-750.

Opposite page top right:
American Electric "Gibson Girl" tandem, c. 1902. Equipped with Burns receiver. Oak, 41" x 9.75" x 11". *Courtesy of Norman M. Mulvey.* $1000-1500.

Opposite page bottom right:
American Electric tandem, c. 1902. Sold by the Keystone Telephone Co. Oak, 48" x 9" x 12". *Courtesy of Norman M. Mulvey.* $500-1000.

This page right:
Blake three box, c. 1890s. Made by DeVeau Telephone Mfg. Co. Walnut, 28.5" x 8" x 6.5". *Courtesy of Norman M. Mulvey.* $500-1000.

Attractive Western Electric "fiddleback," c. 1905. *Courtesy of Art Hyde.* $550-750.

Bottom left:
"Fiddleback," c. 1895. Made by Plummer Ham & Richardson, Worcester, Massachusetts. Oak, 25.75" x 6.25" x 5.75". *Courtesy of Norman M. Mulvey.* $500-750.

Bottom center:
Form of three box telephone, c. 1897. Made by Holtzer Cabot, equipped with Hunning's transmitter. Walnut, 30.5" x 12" x 8". *Courtesy of Norman M. Mulvey.* $800-1200.

Left to right: Ader's system desk subscriber station, French, c. 1882-1889. The transmitter is the flat box on top of the pedestal. Used by the French Co., Le Societe General des Telephones; French cradle telephone, c. 1935. Made by Charles Milde Sons & Co.; L.M. Ericsson wooden wall phone; Sieman's electromagnetic telephone, c. 1878-1884. Siemens & Halske, Berlin.

Used by the Imperial German Post Office as a substitute for the telegraph in rural villages. Served as both transmitter & receiver. Calls were signaled by a whistle in the trumpet. The whistle was removed prior to speaking. *Courtesy of Keith E. Letsche and Dale Martin.* $200-750 ea.

Hand-held telephone used as both transmitter and receiver. Possibly German because of the wood. Blake transmitter. 8.5" long. *Courtesy of Wesley Smith.* Special.

Western Electric lineman's test set, c. 1910s. *Courtesy of Burt Prall.* Special.

Peg system, made by Western Electric. Very rare telephone. *Courtesy of Private Collection.* Special.

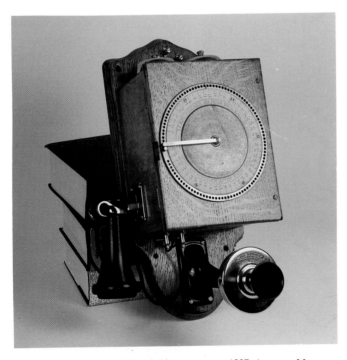

One of the first automatic dial switching systems, c. 1907. Augustus Munson, a Swedish immigrant, started the Munson Automatic Telephone Co. and invented the first automatic dial. Unfortunately, competition defeated him and many did not know of this first dial system. Oak, 16.5" x 8". *Courtesy of Private Collection.* Special.

Chapter Four

Pay Phones

The first pay station was set up in Hartford, Connecticut in 1878. The idea of a telephone operating for a fee was different than the pay phone as we know it today. The first pay stations were elitist booths for those who had money and consisted of a small room furnished with a chair. An attendant would stand at the door to collect the patron's money and would then place their call. These stations were not set up for easy access, instead they gave the utility a sense of formality which attracted the wealthy. However, these pay stations were hard to maintain and were very costly.

3 slot coin collector added to Western Electric tandem, c. 1900. *Courtesy of Norman M. Mulvey.* $1500-2500.

Floor tandem Western Electric silver dollar coin telephone, c. 1894. Gray Telephone Pay Station Co., made for the American Bell Telephone Company. 6'. *Courtesy of Private Collection.* $2000-3500.

In 1888, twelve years after the invention of the first telephone, William Gray, a machinist at Colt's Armory in Hartford, was in dire need of making a telephone call. His wife was critically ill and he desperately needed to call a doctor. When he was denied access to the factory phone, he determined to develop a pay phone for the general public, one which gave everybody equal access to the telephone.

At first he simply did away with the attendant in the booth, and people gained access to the calling room by placing coins in a machine to open the door, placing the call themselves. By the next year, on August 13, 1889, Gray developed the first coin operated telephone which had a receiver that unlocked by the insertion of coins.

Unfortunately for the caller, the lock had to be held down for the duration of the call.

In 1905, Gray developed a pay station with a clear system of coin identification that the operator could hear. His pay station had three elements: top box, with a cranked magneto to get the operator; a funnel shaped transmitter; and a lower box housing the battery jar and coin drop. It used Gray's "resonant" sound connection. Coins inserted in slots fell upon gongs and bells whose distinct sounds were picked up by the transmitter. This enabled the operator to ensure that proper payment was made. If a quarter was dropped into a silver dollar slot, no sound was heard and the operator knew that the correct fee had not been inserted into the phone.

Silver dollar Gray Telephone Pay Station, floor model, c. 1898. Western Electric. Oak, 6' x 16.25" x 17.25". *Courtesy of Norman M. Mulvey.* $2000-3500.

Root coin collector, c. 1895. Used in telephone booths. Mechanical, took only nickels. 7.75" x 6" x 4.5". *Courtesy of Norman M. Mulvey.* $300-350.

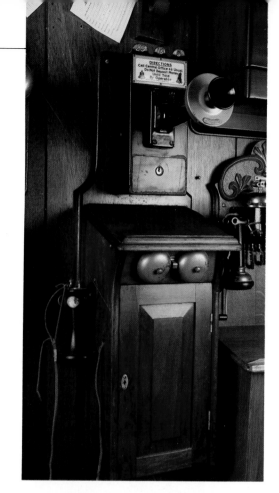

Western Electric model 288 fiddleback with 3 slot coin collector, c. 1898. Made for American Bell. *Courtesy of Norman M. Mulvey.* $550-750.

Gray pay phone, c. 1897. *Courtesy of Vince Wilson.* $350-550.

Bell shaped coin collector on Western Electric 301 fiddleback, c. 1900. Made for Southern New England Telephone (SNET). Walnut. *Courtesy of Norman M. Mulvey.* $1000-1500.

Chicago Telephone Co. pay phone, c. 1900. Oak, 38" x 12". Inside view of coin mechanism. *Courtesy of Norman M. Mulvey.* $800-900.

Silver dollar pay station, c. 1900-1905. 36.5" x 12". *Courtesy of James Goodwin.* $800-1200.

Western Electric fiddleback with 13A coin collector attached, c. early 1900s. Wood, metal, 30" x 9" x 8". *Courtesy of Robert E. Bartlett.* $800-900.

Silver dollar Gray Telephone Pay Station, c. 1904. This pay station was never put into service. Walnut, 37" x 13" x 9". *Courtesy of Norman M. Mulvey.* Special.

Silver dollar Gray Telephone Pay Station, c. 1904. Made for the Kellogg Switchboard & Supply Company. Local battery, walnut, 38.5" x 1.75" x 14". *Courtesy of Norman M. Mulvey.* $400-600.

Fiddleback wall telephone with Model 7J coin collector, c. 1905. Western Electric, used by the Providence Telephone Co. Walnut, 22.75" x 8.25" x 9.25". *Courtesy of Norman M. Mulvey.* $300-550.

Single box wall phone with coin collector, c. 1908. Western Electric. Oak, 19.75" x 9.25" x 5.75". *Courtesy of Norman M. Mulvey.* $300-450.

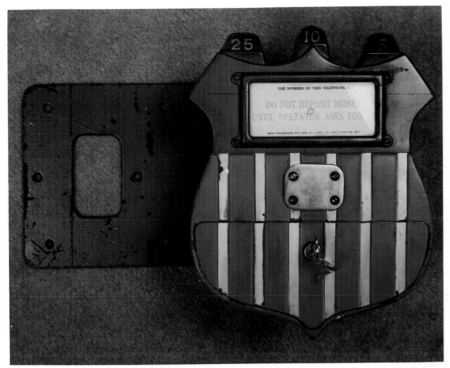

Gray No. 10 shield pay station, patented 1911. 8.5" x 7" x 3.5". Shown on telephone at right. *Courtesy of Private Collection.* Left: $400-600; Right: $650-750.

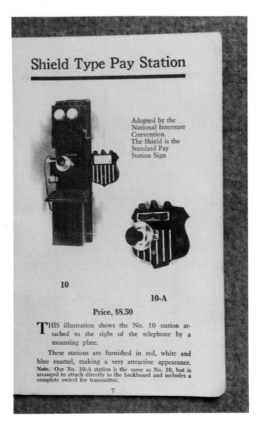

Shield type pay station model no. 10-A, from the Gray Telephone Toll Equipment brochure, 1915. *Courtesy of Odis LeVrier.*

Gray Telephone Pay Station Company, c. 1909. 24.5" x 8.5". *Courtesy of Private Collection.* $250-450.

Gray Telephone pay station attached to a Stromberg-Carlson desk stand, c. 1911. *Courtesy of Private Collection.* $200-375.

Gray silver dollar vanity pay station with sunburst top, equipped with Western Electric components, c. 1910, pictured with owner Odis LeVrier. The nickel, dime, quarter, half-dollar and dollar make different sounds as the coins are dropped in the slots for the operator to differentiate and acknowledge correct payment. *Courtesy of Odis LeVrier.*

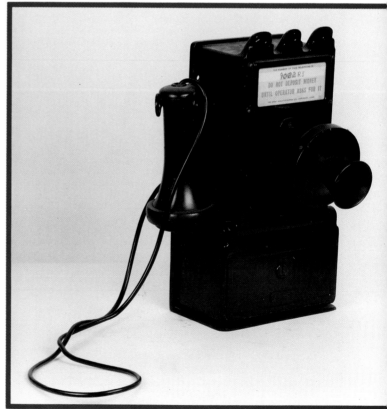

Gray pay station. Patented May 23, 1911. 10.5'' x 6''. *Courtesy of Private Collection.*

Three Gray pay stations, c. May 23, 1911. The pay station in the center was portable. If at a restaurant, it could be brought to your table, pat. Feb. 16. 1909. *Courtesy of Norman M. Mulvey.*
$150-250 ea.

Gray pay station. Patented May 23, 1911. 10.5" x 5.75". *Courtesy of Private Collection.* $275-550.

Gray coin collector with Automatic Electric receiver. 11.5" x 5.75". *Courtesy of Private Collection.* $150-250.

Five variations of the payphone from c. 1910s-1960s. *Courtesy of Burt Prall and Private Collection.* Left: $275-450; Top right: $250-450.

Payphone, model 23J, c. 1940s. The Gray Manufacturing Co., Gray Telephone Pay Station Co., Hartford. Metal, 10.5" x 6". *Courtesy of Robert E. Bartlett.* $175-300.

Portable pay station. 10.25" x 4.5". *Courtesy of Private Collection.* $150-250.

Top left: $125-250; Bottom left: $75-150.

Payphone, c. late 1960s. The 10 button touch tone was introduced in 1964. 17" x 6.5" x 5.5". *Courtesy of Private Collection.* $100-125.

Gray Telephone Pay Station 50A coin collector. Made from 1897-1912. 19" x 6.5". *Courtesy of Private Collection.* $750-1000.

Round calculagraph for measuring long distance calls. Harrison, New Jersey. *Courtesy of Private Collection.*

One of the first credit card payphones, c. 1975. The calls went through the operator so there was no credit card slot. 21". *Courtesy of Wesley Smith.* $50-100.

GTE Automatic Electric model 120-A. Early AE single slot, called a "semi post-pay." Takes money after the party answers. 21" x 6". *Courtesy of Private Collection.* $50-100.

Calculagraph, used to measure time and price of long distance calls. One knob is to start the timer, one to stop it. 7.5'' x 4.25''. *Courtesy of Bob Breish.*

Chapter Five
Intercoms

Intercoms were often set up in businesses and large homes. They were separate from the outside line telephone in that, they were intercommunicators, connecting people within the business or home.

74-station intercom, c. 1897. Clark Automatic Telephone Switchboard Co., Providence, Rhode Island, equipped with **OST** receiver. Very rare telephone. *Courtesy of Private Collection.* $2500-3500.

Tube line intercom, c. 1897. "Rope shaft" with **OST** long pole receiver. Manufacturer unknown. *Courtesy of Private Collection.* $1500-2500.

Manhattan Electric Intercom, c. 1899. Walnut, 5.75" x 5" x 3.5". *Courtesy of Norman M. Mulvey.* $250-350.

"Rope shaft" single button intercom, c. 1898. Equipped with cast iron watchcase receiver, manufacturer unknown, and unmarked Sumter-style transmitter. *Courtesy of Private Collection.* $1200-1500.

Fluted "pencil shaft" intercom, c. 1900. Manufactured by the Manhattan Electric Supply Co. Equipped with a watchcase receiver which is often found on smaller, daintier intercoms of this "pencil shaft" type. The intercom buttons were probably on a separate box. Original nickel with cast iron receiver, wooden base, 10.5" x 3". *Courtesy of Private Collection.* $1500-2500.

Intercoms. Left: wooden desk intercom, pat. Sept. 11, 1906 and March 8, 1910. 8" x 6" x 7.5". Center: probably British, c. 1895, wooden desk set PBX (private branch exchange) intercom. Right: Stromberg-Carleson metal wall intercom. *Courtesy of Barry Erlandson, Keith E. Letshe, and Jerry Williams, consecutively.*
$100-450.

Western Electric intercom, c. 1895. There is no receiver on this painted, black, house intercom with sleigh bells. 4" x 3.5" x 4.75". *Courtesy of Norman M. Mulvey.*

25-station intercom, c. 1901. Marked "Licensed for use in the state of NY only." Automatic Direct Line Telephone Co. 10.5". *Courtesy of Private Collection.* $65-100.

Western Electric upright desk stand, Equipped with watchcase receiver. 12" high. *Courtesy of Private Collection.* $150-300.

Western Electric intercom equipped with watchcase receiver, patented 1904. *Courtesy of Private Collection.* $65-100.

Western Electric upright desk stand intercom, c. 1910. 11.75". *Courtesy of Private Collection.* $75-150.

Western Electric 8-line intercom, c. 1910. Unmarked receiver, American Bell Telephone Co. *Courtesy of Private Collection.* $75-150.

Western Electric upright desk stand intercom, c. 1910. 11.5". *Courtesy of Private Collection.* $75-150.

Upright desk stand intercom, c. 1910. The Connecticut Telephone & Electric Co, with a Western Electric receiver. Connecticut often purchased parts from other large manufacturers and added their own parts. 11.75". *Courtesy of Private Collection.*

Stromberg-Carlson upright desk stand intercom, c. 1920s. 11". *Courtesy of Private Collection.*

Very early Western Electric 205 push button intercom, c. 1927-1939. Brass & steel. *Courtesy of Private Collection.* $125-250.

Circular wooden phone with Ness dial, 7.75" diameter. *Courtesy of Private Collection.* $450-600.

Holtzer-Cabot pedestal box intercom with Ness dial. Equipped with **OST** receiver. 7" x 6" x 4.5". Wooden box, 15". *Courtesy of Private Collection.* $600-1500.

21 station oak wall phone, c. 1910 shown open & closed. Holtzer-Cabot intercom with magneto. 12" x 7" x 6.5". *Courtesy of Private Collection.* $450-600.

Four button desk stand equipped with a Samson transmitter. 10.75". *Courtesy of Private Collection.*

30-station Couch Autophone desk stand, c. 1910. *Courtesy of Private Collection.*

16 dial intercom made by the Holtzer-Cabot Electric Co. 12.5" x 6.5". *Courtesy of Private Collection.* $150-300.

100-station, c. 1910, Auto-phone. Manufactured by The S.H. Couch Co. The rivets in the transmitter faceplate show that this is a Western Electric part made for use by an independent telephone company, such as Couch or Connecticut. *Courtesy of Private Collection.* $100-300.

Left: 25-line intercom. Right: Leofler-Phone 2-line intercom. "Pencil shaft." 12.25". *Courtesy of Private Collection.* $400-600.

Ackerman Boland 6-station intercom, c. 1915. Chicago, equipped with a Federal receiver, 10.75". *Courtesy of Private Collection.* $150-300.

8-line intercom upright desk stand, c. 1910. The labels on the lines indicate that this telephone was probably in use at a theatre. Made by S.H. Couch & Co., Norfolk Downs, Mass. 11.5". *Courtesy of Private Collection.* $100-250.

16-station upright desk stand. Marked, "invented by E.A. Hollis, 1912." 12.25" *Courtesy of Private Collection.* $150-300.

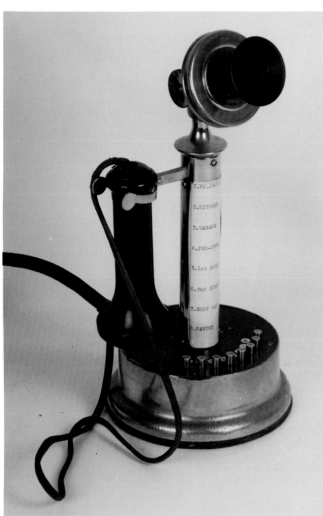

8-station upright desk stand intercom, c. 1904. "Samson Junior" equipped with a **SOLID** receiver, 12". *Courtesy of Private Collection.*

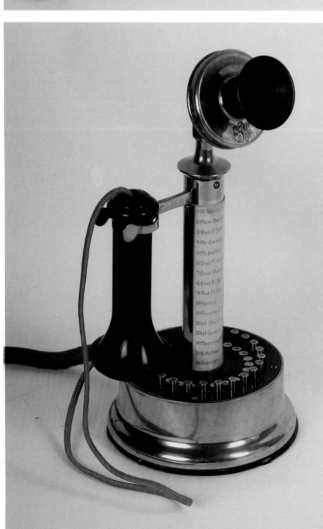

8-station oak intercom, c. 1905. Electrical goods Mfg. Co. with marked transmitter and marked watchcase receiver. 10" x 5.25" x 3". *Courtesy of Private Collection.* $25-75.

16-station intercom, c 1905. Transmitter marked, Electrical Goods Manufacturing Co. "EGM" marked on faceplate. **SOLID** receiver with inside thread mouthpiece. 12.25". *Courtesy of Private Collection.* $100-200.

8-station upright desk stand. Made by Samson Electric Co. equipped with a **SOLID** receiver. 12.25". *Courtesy of Private Collection.* $1000-2000.

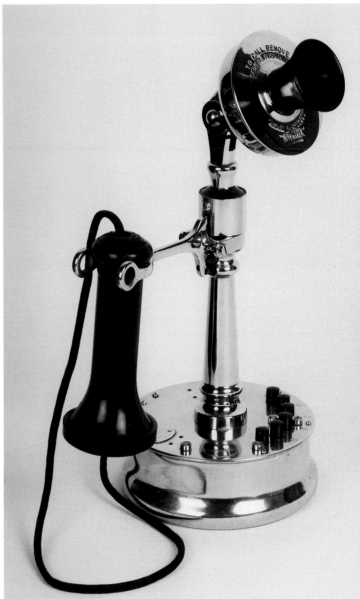

DeVeau "tapered shaft" 10-station intercom, c. 1899-1905. Marked transmitter and **SOLID** receiver. 13". *Courtesy of Private Collection.* $1000-2000.

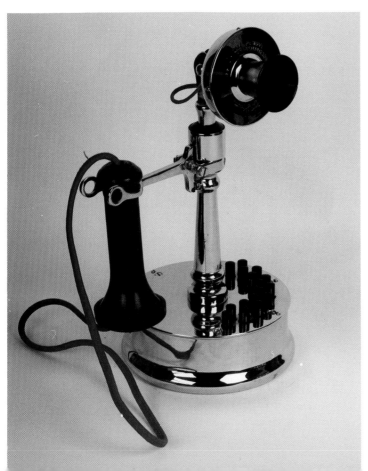

DeVeau 16-station desk stand intercom, c. 1899. DeVeau Telephone Mfg. Co., marked transmitter and receiver tapered shaft. 12.5" x 6.25" diameter. *Courtesy of Private Collection.* $1000-2000.

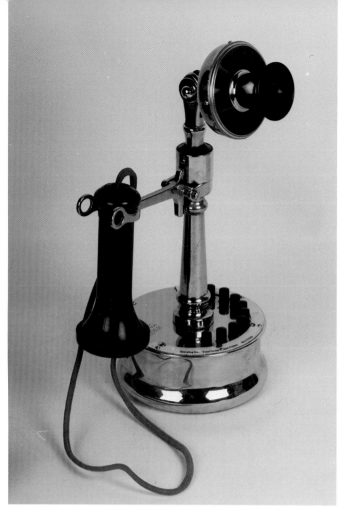

DeVeau 6-station upright desk stand intercom, once used in a hospital, c. 1899. Equipped with marked transmitter and **SOLID** marked receiver. 12.75". *Courtesy of Private Collection.* $550-1000.

Loeffler-Phone 26-station intercom, c. 1897. **OST** receiver and an exposed hook, date this telephone. It was used in a home. Bras, 12" x 8" x 7.5". *Courtesy of Private Collection.* $150-300.

Loeffler 5-button upright desk stand intercom, c. 1900-1905. Early variety of an automatic telephone. Lines indicate owner's room, den, and pantry. Nickel-plated and bakelite, 12" x 5.5". *Courtesy of Ken King.* $300-500.

10-station intercom desk stand, previously used in a bank, c. 1910. Loeffler-Phone equipped with **SOLID** receiver. 11.75". *Courtesy of Private Collection.* $100-150.

Wall intercom equipped with watchcase receiver. 6" x 4.5". *Courtesy of Private Collection.* $25-75.

20-station intercom, c. 1901. Automatic Direct Line Telephone Co. Potbelly design equipped with **OST** receiver. Brass, it used to be nickel-plated, 10" x 5.5". *Courtesy of Private Collection.* $350-500.

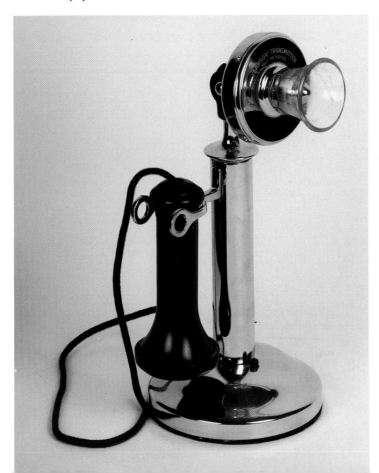

Wooden wall intercom equipped with watchcase receiver. 5" x 5". *Courtesy of Private Collection.* $25-75.

Upright desk stand intercom. Samson transmitter, **SOLID** receiver patented 1900, glass "whisper it" mouthpiece patented Oct., 1916. 11.5". *Courtesy of Private Collection.* $175-250.

33-station Select-O-Phone intercom, c. 1919. Manufactured by the Screw Machine Products Corp., Providence, R.I. *Courtesy of Private Collection.* $100-300.

36-station Select-O-Phone intercom, c. 1919. Manufactured by the Screw Machine Products Corp., Providence, Rhode Island. Equipped with selector base. Calls were made on the Select-O-Phone by turning the dial (using the knob) until the number of the station desired was placed under the indicator. The receiver was then picked up which caused the dial to swing back to its original position and connection was made. Pressing the ringing button rang the bell at the called station. *Courtesy of Private Collection.* $100-300.

Western Electric 4-line intercom, c. 1910. Equipped with watchcase receiver and 302W transmitter. 11.75". *Courtesy of Private Collection.* $100-300.

Intercom from a hospital. Edwards & Co., Norwalk, CT. 5.25" x 6". *Courtesy of Private Collection.* $35-50.

Stromberg-Carlson intercom. Bakelite. *Courtesy of Private Collection.* $25-100.

Commonly called the "mother-in-law" telephone since a second person could listen in (nothing against mother-in-laws of course!). French, equipped with watch-case receiver and grab-a-phone type receiver. 12.5". *Courtesy of Private Collection.* $500-1000.

Stromberg-Carlson intercom. 6.5" x 9". *Courtesy of Private Collection.* $50-125.

Chapter Six
Upright Desk Stands or "Candlesticks"

The upright desk stand, as it is properly called, has in recent years been commonly referred to by its nickname, the "candlestick" telephone. The familiar name refers to the long shaft which resembles a candlestick.

The candlestick telephone consists of a base, a stem, a perch, a mouthpiece, a faceplate, and a cup. The stem varieties include the "fluted shaft," the "tapered shaft," the "roman column" and the "potbelly." These are all nicknames referring to the evident appearances of certain shafts.

The cords also have their own personality. The general types include twisted cords, brown cords, green cords and rattlesnake cords. Original green cords are the most treasured, denoting an older telephone. Reproductions are made, with no intent to deceive, as these reproduction cords are often obvious.

The hook on the candlestick telephone which holds the receiver, also has a varied history. The more pointed the hook, the older it generally is. Hooks invariably graduated from a pointed to a more rounded hook, often with an open circle at the end. Evidently, the more pointed they were, the more easily they broke or damaged the receiver.

Most collector telephones have been renickeled, but collectors prefer to keep the original nickel when still in good quality. These upright desk stands were made of nickel-plating and bakelite. Up until circa 1900, the telephones were nickel plated, and later were bakelite and black painted. Some very early receivers (1870s-1880s) were made of wood, usually oak or walnut.

Various outside terminal (**OST**) receivers representing approximately 95% of the different styles found today. Left to right: unmarked long pole receiver 7.5"; Canadian Bell version of the long pole receiver with 2 red bands, cap marked, 7.5"; Western Electric "pony" receiver 5.5" -6"; American Electric "pony style" receiver 5.75"; American Electric "Burns style" receiver 6.5"; Williams "milk bottle" receiver; Holtzer-Cabot receiver. *Courtesy of Private Collection.* $65-150 ea.

Various inside terminal receivers. Left to right: watchcase receiver (used on an intercom); Swedish-American receiver, 5.5''; Western Electric receiver; SOLID receiver; Federal receiver (rare). *Courtesy of Private Collection.*

Variety of mouthpieces, generally bakelite or hard rubber. Left to right: inside thread mouthpiece, used on the earlier telephones, very desirable among collectors of older phones; standard mouthpiece with outside thread; metal thread mouthpiece, often used on railroad or payphones because it was more durable. *Courtesy of Private Collection.*

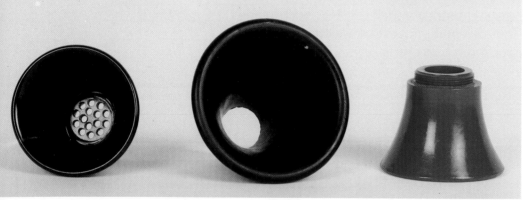

Other types of mouthpieces, left to right: standard size mouthpiece, 1.75'' diameter; larger mouthpiece (rare), 2.5''; later version, colored mouthpiece. *Courtesy of Private Collection.*

The porcelain & glass mouthpieces were considered more sanitary than those made of bakelite or hard rubber. Left to right: porcelain American Electric mouthpiece; "The Dictaphone" glass mouthpiece with metal threads, pat. June 3, 1913; mouthpiece from a grab-a-phone. *Courtesy of Private Collection.*

Three standard mouthpieces with advertising. The far right mouthpiece is porcelain. *Courtesy of Private Collection.*

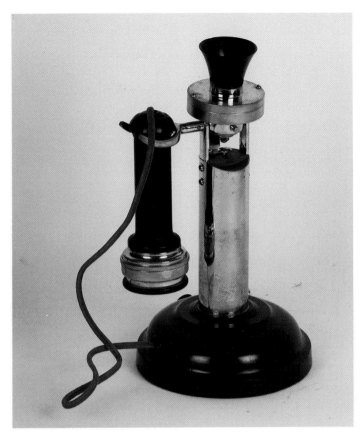

Stromberg-Carlson upright desk stand, c. 1896. Swivel top and pointed hook switch which denotes an older telephone. Very rare. *Courtesy of Private Collection.* $2000-4000.

Roman column upright desk stand, c. 1897. Same base as the previous year's model. A very desirable telephone for collectors. *Courtesy of Private Collection.* $1500-2200.

Stromberg-Carlson "oil can," c. 1898. "Oil can" is a nick-name for this type of S-C because of the base's resemblance to an old oil can. Notice the hook switch for the receiver is now circular rather than pointed as in the older phone above. Perhaps they changed to the circular because it was damaged less easily as people hung up the receiver. Cast brass base, 11.5". *Courtesy of Private Collection.* $375-600.

Stromberg-Carlson introduced this model c. 1904. It is often referred to as the "Kansas City" type candlestick because advertisements for the phone depicted Kansas City. Glass mouthpiece, c. 1900, embossed transmitter. 11.5". *Courtesy of Private Collection.* $350-550.

Stromberg-Carlson "oil can," c. 1899. This model was first produced in 1898 and was made for approximately 8 years. Notice the ball behind the transmitter. Renickeled, 12". *Courtesy of Ken King.* $375-600.

Strowger wall phone displaying Chinese characters on the dial. It probably originated in the San Francisco area, c. 1908. 8.75" x 7". *Courtesy of Private Collection.* $600-750.

Last model of the Stromberg-Carlson "oil can," c. 1900-1905. In this model they eliminated the ball behind the transmitter and used a larger transmitter. 11.5". *Courtesy of Private Collection.* $350-550.

The next model of the Stromberg-Carlson had a black base. This is perhaps the last model of "candlestick" introduced by S-C, c. 1905. 11.75". Sitting on top of wooden box is a salesman's sample. *Barry Erlandson & Ken King.* Salesman's sample: $85-125, box $35-55; right: $100-150.

Stromberg-Carlson dial upright desk stand, c. mid-1920s. Used Automatic Electric dial until about WWII (marked Strowger). 11.75". *Courtesy of Jim Aita.* $100-200.

Stromberg-Carlson Corporation Advertisement "One hundred Years of Excellence." *Courtesy of Odis LeVrier.*

Square metal case Strowger wall set, c. 1904. The pointed switch hook denotes an older telephone. *Courtesy of Private Collection.* $800-1500.

Automatic Electric 10-hole dial wheel, c. early 1903. The first practical upright desk stand, it has no push button to ring the other end, and displays a "O" for operator, as today. 13". Rare. *Courtesy of Private Collection.* $3500-5000.

The second model of the Strowger dial telephone shows an 11-hole dial, c. early 1904. Both the long-distance and the O were wired together, unknown to the public. *Courtesy of Private Collection.* $1500-2500.

The next model of the Strowger desk stand shows changed in the size of the transmitter and a new feature of the "A" on the switch hook, c. 1905. The porcelain mouthpiece was preferred in the sanitation-conscious days of the early 1900s because it was thought to be germ free. *Courtesy of Private Collection.*

Strowger wall phone, c. 1907-1908. It had the same component parts as the dial set, but was made to hang on the wall. *Courtesy of Private Collection.* $500-750.

American Electric "fluted shaft" desk stand, c. 1900. Marked transmitter and marked **OST** pony receiver. Cast iron base, 11". *Courtesy of Private Collection.* $1500-2500.

Two Automatic Electric dial telephones, c. 1913-1919. Left: has a sunburst dial, so named because of the decoration on the dial. This telephone was only made from 1904-1913 because of mechanical problems. The wheel does not retract, it stays at the number dialed. Right: painted black Automatic Electric Co. upright desk stand. *Courtesy of Private Collection.* Left: $300-400; Right: $200-300.

Siemens Brothers & Co., Ltd. telephone with Strowger dial, c. 1905-1909. Made in Germany, although the company had a factory in Woolwich, Kentucky. It is believed that siemens products were sold only to German customers but used Strowger equipment. 10" x 10.5". *Courtesy of Private Collection.* $1500-2000.

Automatic Electric "straight shaft" upright desk stand, c. 1912. Equipped with a "Wonderphone" transmitter and marked receiver. Cast iron base, 11.5". *Courtesy of Private Collection.* $250-400.

Decorative "Eiffel Tower" desk set, c. 1900. L.M. Ericsson. 12" x 10". *Courtesy of Private Collection.* $650-1250.

Article on Siemens & Brothers Co., Ltd. from *Automatic Telephony. Courtesy of Private Collection.*

"Eiffel Tower" desk set, c. 1890s. L.M. Ericson & Co. *Courtesy of Roger Natte.*

1897 "San Francisco" model, Western Electric with glass Red Cross mouthpiece to show the mouthpiece is more sanitary. American Bell marked transmitter and **OST** "pony" receiver. Patented '89, '91, '92. *Courtesy of Private Collection.* $2500-3500.

L.M. Ericsson desk set, c. 1905. Marked, "Wein" "Decker & Homolka." *Courtesy of Private Collection.* $100-300.

Western Electric "potbelly" with telescopic stem, c. 1895. Beveled edge transmitter and porcelain mouthpiece, marked long pole **OST** receiver, forked hook. 12.25". *Courtesy of Private Collection.* $4000-5000.

Western Electric "potbelly," c. 1897 pictured at full extension. **OST** long pole receiver with telescopic stem. 14.75" x 6". *Courtesy of Private Collection.* $3500-5000.

Whitman & Couch "potbelly," c. 1897. Forked hook and original **OST** Couch receiver. 11.5" x 4.75". *Courtesy of Private Collection.* $1500-2500.

Western Electric "potbelly." Equipped with **OST** long pole receiver and whisper-it mouthpiece. 12" x 6" x 5" diameter. *Courtesy of Odis LeVrier.* $2500-3500.

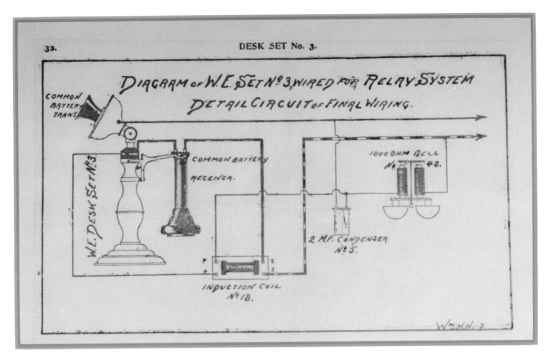

Diagram of Western Electric set no. 3. *Courtesy of Odis LeVrier.*

The North Electric Company "potbelly," c. 1901. Marked transmitter and receiver with forked switch hook. 11" x 5.5". *Courtesy of Private Collection.* $1200-1800.

Chicago Genuine Solid Back "potbelly," Chicago Telephone Supply Co., c. 1898. Notice that the exposed wire connected to the transmitter on the phones above is now concealed inside, c. 1900. *Courtesy of Private Collection.* $1200-1800.

* Page 96&97: All $1200-2500 ea.

Acme Electric Co., c. 1902. Equipped with OST receiver. 10 1/2'' x 5''. *Courtesy of Private Collection.*

Utica Fire Alarm Telephone Co., c. 1900. Equipped with OST receiver. 11'' x 5.5''. *Courtesy of Private Collection*

''Potbelly'' c. 1902. Mfg. by Couch & Seeley, 11.5'' x 4.5''. *Courtesy of Private Collection.*

Swedish-American "potbelly," c. 1901. Equipped with porcelain mouthpiece, OST receiver, and beveled edge transmitter. Exposed wire into transmitter. 10.5" x 5.5". *Courtesy of Private Collection.*

Connecticut "potbelly," c. 1900. Equipped with SOLID OST receiver, closed receiver hook, inside thread mouthpiece. 10.5" x 5". *Courtesy of Private Collection.*

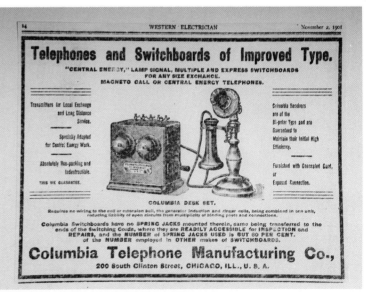

Elliott Telephone & Electric Co., Indianapolis, c. 1901. "Potbelly" with OST receiver. 11" x 5". *Courtesy of Private Collection.*

Advertisement from the *Western Electrician.*, Nov. 2, 1901. *Courtesy of Odis LeVrier.*

American Electric "potbelly," c. 1902. **OST** Burns receiver with marked transmitter. *Courtesy of Private Collection.* $1000-1500.

The same "semi-potbelly" as above with a transmitter marked Montgomery Ward & Co., c. 1905. *Courtesy of Private Collection.* $300-750.

Genuine solid black "semi-potbelly," c. 1905. Chicago Telephone Supply Co. 11". *Courtesy of Private Collection.* $300-500.

Very attractive tall "potbelly" desk stand, c. 1896. Wilhelm Telephone Mfg. Co., (extremely rare). Equipped with unique transmitter, larger than usual, **OST** receiver, molded mouthpiece and push button for operator perhaps. 13.25" x 6". *Courtesy of Private Collection.* $2500-5000.

B-R Electric pedestal model, c. 1902. Unusual telephone with the circular part in the stem above the potbelly, **OST** receiver. 14.5" x 5". *Courtesy of Private Collection.* $1000-1500.

American Bell pedestal "potbelly" with **OST** receiver. 15.5". *Courtesy of Private Collection.* $1500-2500.

An unusual black Western Electric No. 10, c. 1900. **OST** receiver. 12". *Courtesy of Private Collection.* $1000-1500.

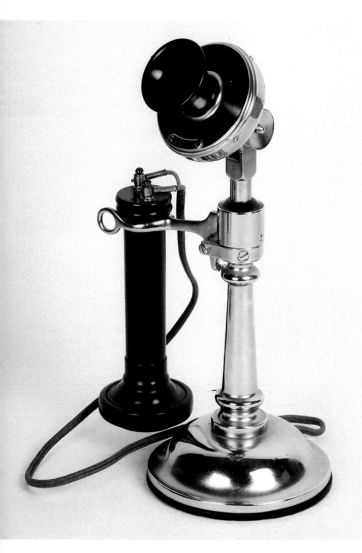

Western Electric No. 10 style desk stand, c. 1898. **OST** receiver. 12". *Courtesy of Private Collection.* $1200-1500.

Northern Electric (Canadian Bell) No. 10 style desk stand. **OST** receiver. 12.5". *Courtesy of Private Collection.* $1000-1200.

The Williams Telephone & Supply Co., "tapered shaft" desk stand, c. 1900. Equipped with "milk bottle" **OST** receiver and reproduction green cord. *Courtesy of Private Collection.* $1500-2000.

"Tapered shaft" desk stand, c. 1902. Equipped with B-R marked transmitter, marked **SOLID OST** receiver, and closed ring hook. 11.25" x 5". *Courtesy of Private Collection.* $800-1200.

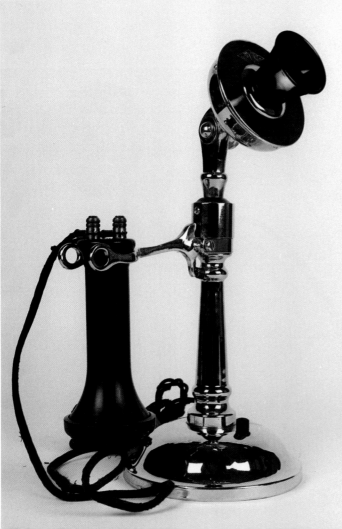

DeVeau Telephone Mfg. Co., c. 1901. **SOLID OST** receiver with single push-button, probably a signaling button. 12" x 6.5". *Courtesy of Private Collection.* $650-850.

Eastern Telephone Co. "tapered" desk stand, c. 1900-1905. 12" x 5.5". *Courtesy of Ken King.* $600-800.

B-R Electric & Tel. Mfg. Co., c. 1901. Equipped with Kansas City Telephone Mfg. Co. transmitter on a B-R body. Long **OST** receiver, forked hook. 11" x 5". *Courtesy of Private Collection.* $800-1000.

"Tapered shaft" desk stand, c. 1901. Marked International, Chicago on transmitter 10.25" x 5.5". *Courtesy of Private Collection.* $550-850.

Eureka Electric Co. "tapered shaft," c. 1902. 11.25". *Courtesy of Odis LeVrier.* $1500-2500.

Manhattan Electric Supply Co. "tapered shaft," c. 1903. Beveled edge transmitter with hollow mouthpiece—no holes. Exposed wire on opposite side. 11". *Courtesy of Private Collection.* $1700-2500.

Monarch Telephone Mfg. Co., c. 1904. Early marked transmitter & heavy cast iron base. 11.5". *Courtesy of Private Collection.* $200-400.

Sterling Electric Co. upright desk stand, c. 1904. Sterling bell base with marked Sterling receiver (unusual). *Courtesy of Private Collection.* $1500-2500.

Stromberg-Carlson upright desk stand, c. 1905 with attachments for pencil on mouthpiece and a notepad on base of stem. 12". *Courtesy of Norman Mulvey.* $100-250.

B-R desk stand, c. 1906. This style of telephone is often found with a different manufacturing company's transmitter & marketed under their name, not B-R's name. At least 10 different companies have been identified. 10". *Courtesy of Private Collection.* $200-300.

Kellogg Switchboard & Supply Co. upright desk stand, c. 1901. Kellogg early transmitter & receiver, beveled edge transmitter, metal thread mouthpiece (heavy duty often used for railroad or paystations). 10.5". *Courtesy of Private Collection.* $300-400.

Example of Julius Andrae & Sons Co. using the upright desk stand above with their own lettered transmitter, pat. 1902. Other companies also used their transmitters on this same set. *Courtesy of Private Collection.* $200-450.

American Electric, c. 1912. For use by the Keystone Telephone Co. Painted black. *Courtesy of Private Collection.* $150-250.

Swedish-American "straight shaft," c. 1912. It has a steel body, as opposed to brass, and therefore is not nickel-plated since nickel did not adhere well to steel. Flat transmitter faceplate came after the beveled edge transmitter. 10.5". *Courtesy of Private Collection.* $150-250.

Swedish-American desk stand, c. 1901. Not a rare phone but attractive, model #70. **OST** receiver, porcelain mouthpiece, original wooden base, beveled edge transmitter 12". *Courtesy of Private Collection.* $400-500.

Williams-Abbott Electric Co., upright desk stand, c. 1904. Beveled edge transmitter, unmarked Williams-Abbott receiver. 10.5". *Courtesy of Private Collection.* $500-700.

The Sumter Telephone Manufacturing Co. 10.5". *Courtesy of Private Collection.* $700-1000.

Holtzer-Cabot, Boston, desk stand. **OST** receiver & non-adjustable transmitter 10.75". *Courtesy of Private Collection.* $500-800.

The Sumter Telephone Mfg. Co. Outside wire to transmitter. 10.5". *Courtesy of Private Collection.* $700-1000.

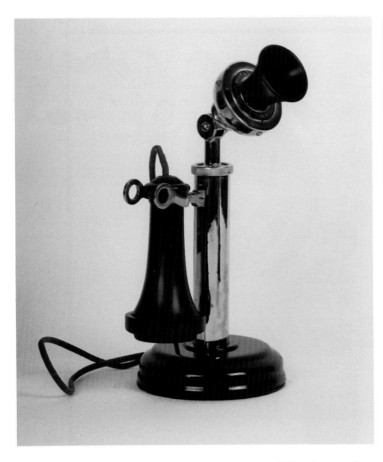

The Sumter Telephone Mfg. Co. c. pre-1910. Equipped with flat edge transmitter. 11". *Courtesy of Private Collection.* $800-1000.

The Century Telephone Construction Co., c. 1912. The tag is on the base and it has an inside thread mouthpiece. *Courtesy of Private Collection.* $150-250.

The Sumter Telephone Mfg. Co., all-black desk stand c. 1910. 9.5". *Courtesy of Private Collection.* $450-650.

Rare Century Telephone Construction Co. upright desk stand, c. 1900. Forked hook, marked transmitter, **OST** "pony" receiver, "Whispering" glass mouthpiece patented Oct. 1916, original cor. 11.5". *Courtesy of Private Collection.* $2000-3500.

"Thick shaft" desk stand, c. 1912. North Electric Co. Flat cast iron base, marked transmitter & receiver. 9.5". *Courtesy of Private Collection.* $250-450.

Monarch Telephone Manufacturing Co., c. 1914. Fort Dodge, Iowa. Black trimmed in nickel. *Courtesy of Private Collection.* $100-250.

Chicago Genuine Solid Back, c. 1915. Chicago Telephone Supply Co., Elkhardt, Indian. Painted black and trimmed in nickel. *Courtesy of Private Collection.* $150-250.

"Wonderphone," c. 1915. Cracraft-Leich Electric Co. with "Wonderphone" transmitter. 11.5". *Courtesy of Private Collection.* $250-350.

All black upright desk stand, c. 1910. Dean Electric Telephone Apparatus Co. 10.5". *Courtesy of Private Collection.* $125-225.

Federal Telephone & Telegraph Co., Buffalo, New York, c. 1919. Unusually short Federal receiver with unmarked transmitter & receiver. Marked base. *Courtesy of Private Collection.* $175-275.

Two Western Electric upright desk stands with different transmitters. Left: flat transmitter, marked Property of American Bell Telephone Co., circa 1904. Right: beveled edge transmitter pat'd. '92, **OST** receiver, outside wire to transmitter. *Courtesy of Private Collection.* $150-350.

Canadian Bell desk stand. Transmitter marked, "Property of the Bell Telephone Company of Canada Limited." Equipped with **OST** double red band receiver. 11.5". *Courtesy of Private Collection.* $1200-2000.

Another Western Electric upright desk stand, c. 1900. Marked **OST** receiver. 11.25". Rare. *Courtesy of Private Collection.* $2000-4000.

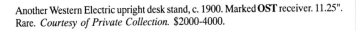

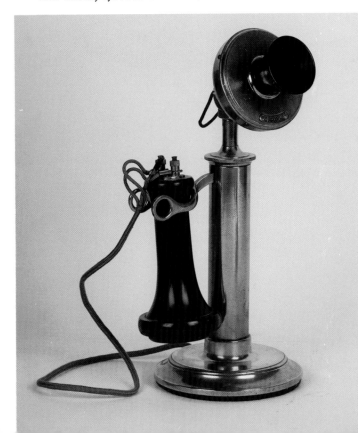

Two upright desk stands appear to have the same body with a different company's transmitters. Left: c. 1906, "Samson Jr." transmitter pat'd. Feb 10, 1903. **SOLID** receiver 11.75". Right: The Williams Telephone & Supply Co. transmitter, **SOLID** receiver. 11.75". *Courtesy of Private Collection.* $200-400.

Western Electric left-handed desk stand, c. 1920. Factory-made to accommodate a receiver on the right side instead of on the left as is usual. Drop receiver hook. 12". *Courtesy of Private Collection.* $250-350.

American Electric "straight shaft," c. 1902. Equipped with adjustable transmitter, porcelain mouthpiece, **OST** marked receiver. 11.5". *Courtesy of Private Collection.* $250-350.

Western Electric upright desk stand. Unmarked receiver with unusual mouthpiece. 11.5". *Courtesy of Private Collection.* $250-350.

American Electric "straight shaft," c. 1903. Non-adjustable transmitter. 11". *Courtesy of Private Collection.* $150-250.

Three attractive upright desk stands. Left: Beeman's "rope shaft," brass, original nickel plating, wood base, c. mid-1890s. Center: "Roman column" manufactured by Stromberg-Carlson, c. 1897. Right: Chicago Hotel Tel. Co. wooden, and nickel "potbelly," c. mid-1890s. *Courtesy of Paul McFadden.* Left-right: $2000-3000; $1500-2500; $2500-4000.

L.M. Ericson, England, c. 1920s. Manufactured in Plymouth, England, the plant was destroyed in WWII. Later version transmitter with all-brass receiver. 12.5". *Courtesy of Private Collection.* $150-250.

Northern Electric black dial set. Marked transmitter & receiver. *Courtesy of Private Collection.* $175-200.

Western Electric black dial set is the same as the Canadian & English versions. Made from c. 1919-1941. 12". *Courtesy of Private Collection.* $150-225.

Western Electric black dial upright desk stand, c. 1920s-1930s. Porcelain advertising mouthpiece, brass painted black (some are steel base & stem, test whether steel or brass by using a magnet, if brass, the magnet will not attract). 12". *Courtesy of Private Collection.* $150-300.

Western Electric railroad dial upright desk stand, c. 1920s-1930s. Generally used for railroad communications. Kellogg & Western Electric had the largest quantities in circulation. *Courtesy of Private Collection.* $150-200.

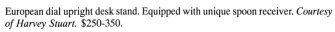

European dial upright desk stand. Equipped with unique spoon receiver. *Courtesy of Harvey Stuart.* $250-350.

Scissor gates with Western Electric desk stand, c. 1907. Scissor gates were mounted to desks with the ability to expand and detract for easy mobility. Pencil attachment. 20" extension. *Courtesy of Norman Mulvey.* $200-250.

Small fluted "pencil shaft" desk stand, c. 1900. Manufactured by the Elston Telephone Co. Equipped with **OST** receiver & marked Elston transmitter faceplate. 10". *Courtesy of Private Collection.* $1500-3000.

"Fluted shaft" upright desk stand, c. 1902. Manufactured by the Sterling Electric Co., Lafayette, IN. 12.5". *Courtesy of Private Collection.* $2000-3000.

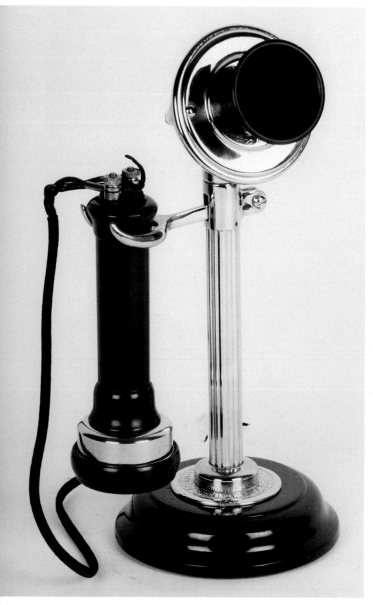

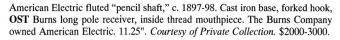

American Electric fluted "pencil shaft," c. 1897-98. Cast iron base, forked hook, **OST** Burns long pole receiver, inside thread mouthpiece. The Burns Company owned American Electric. 11.25". *Courtesy of Private Collection.* $2000-3000.

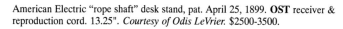

American Electric "rope shaft" desk stand, pat. April 25, 1899. **OST** receiver & reproduction cord. 13.25". *Courtesy of Odis LeVrier.* $2500-3500.

Fancy upright desk stand, 10.25". Rare. *Courtesy of Private Collection.* Very rare.

Fancy upright desk stand. Lockwood Co. Only two telephones known to exist by this company. Marble bottom. 11.25". *Courtesy of Private Collection.* Very rare.

Called the "golf ball" telephone, a British version of the Western Electric No. 10 style desk stand. 12". Rare. *Courtesy of Private Collection.* $2500-4000.

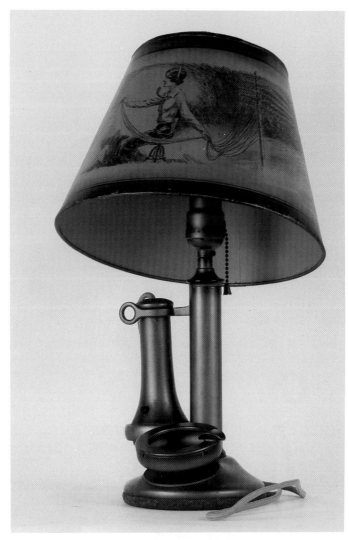

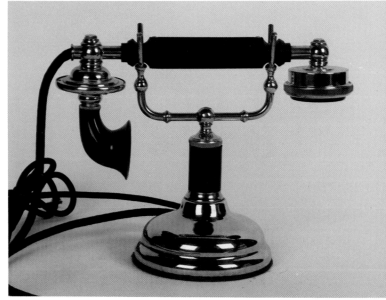

Grab-a-phone, c. 1915-1925. Federal Tel. & Tel. Co., Buffalo, New York. 8" x 10". *Courtesy of Private Collection.* $150-250.

Western Electric desk stand transformed into a lamp, ashtray & lighter. Pioneers of America fundraiser. *Courtesy of Paul McFadden.* $75-100.

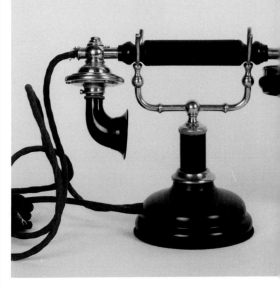

Upright desk stand made especially for Mary Pickford, famous movie star in the early 1900s. 11.25". *Courtesy of Private Collection.* Very rare.

European desk set grab-a-phone, c. 1900. Unmarked. Wood & nickel. *Courtesy of Private Collection.* $200-300.

Microhand telephones. Left: L.M. Ericsson Telephone Mfg. Co., Buffalo, New York. 8" x 10". $150-200. Right: Kellogg Telephone Co., 7" x 9.5". *Courtesy of Private Collection.* $200-300.

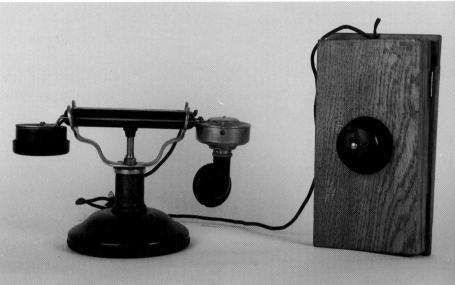

Microhand telephone, c. 1915. DeVeau Telephone Mfg. Co. *Courtesy of Paul McFadden.* $200-300.

Chapter Seven

Cradle Telephones and Color Phones

The first Western Electric handset telephone, model AA1, "the chopped off dial candlestick with a cradle," first introduced c. 1926. It was thought to be manufactured only in black, however, a gaudy gold and a silver version are known. It was only available for one year making this a very rare telephone. Metal base, bakelite handset, 5.75" x 9". *Courtesy of Mike Irvin.* $250-550.

Eventually the telephone became more compact and therefore more facilitative for the public. The "candlestick" telephone was cut down into what some call the "cradle" telephone, the telephone as we are familiar with it today. The progression began with the AA1 model, through the models 102, 202, 300, 3500, and the 500. Soon after, the Princess™ telephone was introduced and then the even more compact Trimline™ became popular.

With the advent of the 300 type desk set in 1939 from Western Electric, an innovation entered the industry which would alter the telephone in wondrous ways. The bell was placed in the base of the model, instead of in a separate box. Such a novel idea spread quickly. The early versions were made of metal, but plastic was substituted in the 1940s.

The phone in the 1950s became a haven for teens in America. The phone was the means by which the teen-ager could gossip and escape the life at home without actually leaving home. Popular novels and plays reflected this occurrence. J.D. Salinger's 1951 best-selling novel, *The Catcher in the Rye* and the musical *Bye Bye Birdie*, are both examples which depict the telephone as a popular device for the teenager.

Western Electric AA1 round base. *Courtesy of Private Collection.* $300-550.

Western Electric flat wall telephone. This telephone is thought to have been introduced roughly the same time as the AA1. 12" long. *Courtesy of Gerald Gapa.* $125-250.

Western Electric model 102, was available from c. 1928-1929. 5.5" x 8.5". *Courtesy of Charles Stanley.* $100-300.

Left: Western Electric oval base model 292, c. 1930. Right: Western Electric round base model 102. *Courtesy of Private Collection.* $75-200 ea.

Western Electric round base model 102 with Hush-A-Phone attached to receiver. Hush-A-Phone Corp., New York. 6" x 11". *Courtesy of Private Collection.*

Three Western Electric cradle telephone. Left: model AA1, c. 1927, brass; center: model 102, c. 1929, potmetal; right: model 202, c. 1930-1937, potmetal. *Courtesy of Mike Irvin and George Howard.* $100-500 ea.

Olive round base Western Electric model 102, black oval base Western Electric model 202. *Courtesy of Private Collection.* $100-200.

Red and ivory Western Electric models 202, c. 1935. *Courtesy of Private Collection.*

Green set of Western Electric models 202, familiarly known as the "continental." *Courtesy of Private Collection.*

* Page 123: All $150-250 ea.

Western Electric model 202 with ringer, c. 1930. 8.5" x 9.5" wood box. *Courtesy of Jerry Williams.*

Western Electric gold model 202, c. 1931. Metal. *Courtesy of Charles Stanley.* $100-200.

Silver & ivory, gold & ivory Western Electric model 202, also known as the "Imperial." Produced in 1951 as special 75th Anniversary sets. *Courtesy of Private Collection.* $85-175 ea.

Black Western Electric variation model 205, c. 1938. *Courtesy of Private Collection.* $150-300.

page 125

Western Electric Co. P.B.X. systems desk set, No. 750-A, 1929. *Courtesy of Private Collection.* $150-250.

Advertisement, Saturday Evening Post, November 6, 1943. *Courtesy of Gerry Gapa.*

Western Electric rotary dial desk set, prototype set mounted on a subset, c. 1951. 6.25" x 7". *Courtesy of Private Collection.* $75-150.

Two Western Electric 300 model type desk sets, c. 1937. Bells were placed in the base of the telephone, a new idea! *Courtesy of Private Collection.* $35-175 ea.

Blue and green versions of the Western Electric 300 type models. *Courtesy of Private Collection.* $150-500 each.

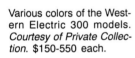

Various colors of the Western Electric 300 models. *Courtesy of Private Collection.* $150-550 each.

Ivory Western Electric model 304 desk set, c. 1950s. *Courtesy of Private Collection.* $100-175.

Ivory Western Electric model 354 wall set with F1 handset. The 354 was manufactured from 1950-1954. *Courtesy of Private Collection.* $50-200.

Western Electric advertisement of the eight colors available. *Courtesy of Gerry Gapa.*

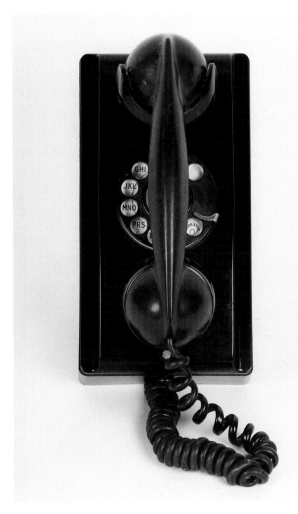

Western Electric black model 354 wall set with F1 handset, 1950-1954. *Courtesy of Private Collection.* $50-150.

The body changed from the Western Electric 300 type model desk set to the 500 set shown at right, in the late 1950s. *Courtesy of Private Collection.* $25-60 ea.

"Western Electric 500 set. *Courtesy of Private Collection.* $25-50.

Northern Electric Co. wall set, pat. 1922, 23, 35. *Courtesy of Private Collection.* $20-35.

Western Electric 500 clear plastic desk set. *Courtesy of Private Collection.* $85-150.

Western Electric demonstration set. *Courtesy of Private Collection.*

Cute advertisement showing the wonders of the telephone. *Courtesy of Gerald Gapa.*

Black Western Electric with Lucite buttons, c. 1960s. *Courtesy of Private Collection. $45-75.*

Wonderful advertisement showing the colors available in the new Western Electric 500 sets. *Courtesy of Gerald Gapa.*

Wonderful display of Western Electric's colored 202 (top row), 302 (middle row), 5302 (a cross between the 302 and the 500), and the 500 set. Plastic cradle phones were introduced c. 1949-1952. The 500 set came out c. 1949. *Courtesy of Jim Aita.*

"Princess" telephones, c. 1959. 8" x 3.5". *Courtesy of Private Collection.* $18-55 ea.

Western Electric prototype telephone, c. 1957-58. *Courtesy of Gerald Gapa.*

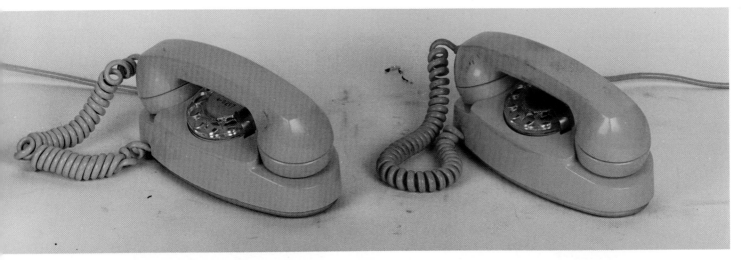

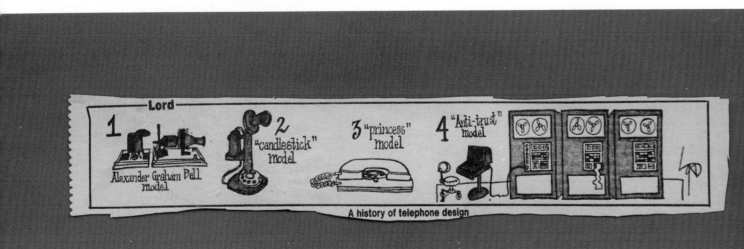

Cartoon depicting the monopoly of the Bell System. *Courtesy of Gerald Gapa.*

First automatic dial phone, c. early 1960s. Originally used in the military, these card dialers later became available to the general public. *Courtesy of Bob Breish, cards courtesy of Wesley Smith.*

November 7, 1959 advertisement for Western Electric. *Courtesy of Gerald Gapa.* $3-10.

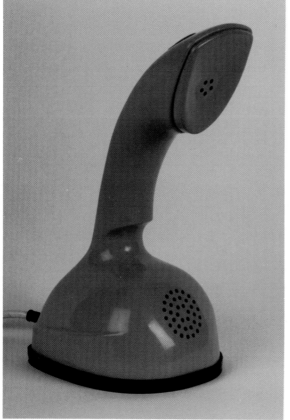

"Erica" phone, similar phone first produced by Ericsson, thereby its nickname. North Electric, c. 1960s. *Courtesy of Larry Garnatz.* $20-65.

Push button, c. 1964. *Courtesy of Private Collection.* $5-25.

Southwestern Bell advertisement for the new "Trimline" telephone. *Courtesy of Private Collection.*

North Electric advertisement showing the telephone as a resource for the teenager during the 1960s. *Courtesy of Private Collection.*

Western Electric "Trimlines" c. 1968. *Courtesy of Private Collection.* $10-35 ea.

Rotary dial and push button "trimline" telephones, c.1968. *Courtesy of Private Collection.*

Western Electric "One in a million" telephone. One in every million telephones were plated in gold and given to an employee of the company. This one in a million was made in 1969. *Courtesy of Mike Irvin.* Rare.

Automatic Electric "monophone" c. 1928. 5.5" x 9". *Courtesy of Charles Stanley.* $50-100.

Automatic Electric 2-line "monophone," c. 1940s. *Courtesy of Private Collection.* $100-150.

Art Deco Automatic Electric "monophone" with dial, c. 1928. Bakelite, 5" x 9". *Courtesy of Charles Stanley.* $100-200.

Automatic Electric with ringer inside, c. late 1930s-early '40s. *Courtesy of Private Collection.* $125-200.

Automatic Electric "monophone" round base in maroon. *Courtesy of Private Collection.* $125-200.

Automatic Electric with square base to fit ringer. 6.75" x 8" x 5.25". *Courtesy of Private Collection.* $100-150.

Various colors of the Automatic Electric desk sets type 34A. Some collectors call this the "Shirley Temple" phone. $200-450 ea.

Automatic Electric "monophone" type 34A intercom. First introduced in 1934. Base, 8.25" deep, 5.25" wide. *Courtesy of Private Collection.* $100-150.

Automatic Electric wall phone type 35A. It was first introduced in 1935 with a new "all position" transmitter. *Courtesy of Private Collection.* $100-200.

Desk crank and wall crank sets. Assembled during the war with salvaged pieces using Leich handsets. *Courtesy of Charles and Bill Stanley.* $85-125 ea.

Beehive desk/wall crank, c. 1930s. Leich Electric. One of the last crank telephones used. Bakelite, 10" x 5.5". *Courtesy of Jim Aita.* $35-50.

Leich Electric desk set, c. 1935. Greenish-clear plastic, 5.75" x 7". *Courtesy of Charles Stanley.* $150-300.

Leich Electric desk set. Bakelite, 6.5" x 5" x 8". *Courtesy of Charles Stanley.* $25-50.

Rare orange colored Leich Electric desk set, c. 1935. Bakelite, 5.57" x 9". *Courtesy of Charles Stanley.* $150-300.

Stromberg-Carlson magneto desk set, c. 1940s. This telephone gave the option of crank or dial. Bakelite, 8'' x 8''. *Courtesy of Paul McFadden.*

Stromberg-Carlson clear plastic desk set, c. 1930s. Model #1212 AB. 5.5" x 9". *Courtesy of G.K. Hillestad.* $200-300.

Connecticut Telephone & Electric Co. desk set, c. 1949. TP-6-A. Metal, 6" x 8". *Courtesy of Charles Stanley.* $25-75.

Kellogg Switchboard & Supply Co., "Masterphone" non-dial desk set. 4/25" x 9.5". *Courtesy of Private Collection.* $75-100.

Explosion-proof telephone, c. 1940s. Equipped with 302 handset. This type of telephone was to keep electrical sparks from lighting in dangerous areas, used in old mines and refinery stations. *Courtesy of Paul McFadden.* $50-100.

Kellogg "masterphone" dial desk set also referred to as the "ash tray phone," c. late 1920s-30s. *Courtesy of Private Collection.* $75-125.

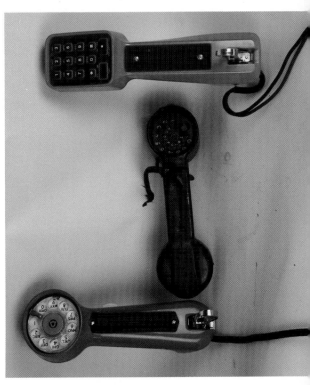

Lineman's test sets. *Courtesy of Burt Prall.* $10-35.

Kellogg "masterphone 900," c. 1928. Bakelite, 5.5" x 9". *Courtesy of Charles Stanley.* $35-95.

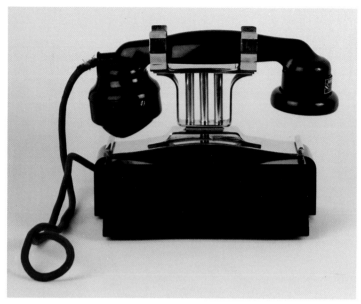

French desk set, maker unknown. *Courtesy of Wesley Smith.* $200-300.

Round base dial desk set. *Courtesy of Private Collection.* $50-100.

Ericsson desk set with one button. Made in Holland, 5" x 6" x 7.5". *Courtesy of Private Collection.* $25-65.

A decorative type of cradle phone certainly, this teapot was supposedly once owned by an Arab Sheik. 17" high. *Courtesy of Zora Notanblut.* $100-200.

Chapter Eight
Novelty Telephones

Alexander Bell may never have forseen what his invention would come to, but in the 1960s, 70s, and 80s, the telephone superceded its main function as a communicative device and took on a new life as a form of art.

The Snoopy Woodstock telephone. American Telecommunications Corp., El Monte, Calif. 7.5" x 14" x 8". *Courtesy of Private Collection.* $65-125.

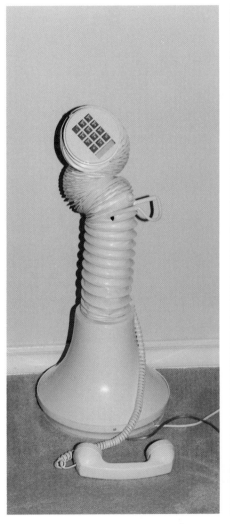

Telephone constricted in use and extended when not in use. Plastic. *Courtesy of Private Collection.* $50-150.

Kermit telephone. American Telecommunications Corp., 1983 Henson Assoc. Kermit reclines in his seat as you pick up the receiver. 11.5" x 8.5" x 8". *Courtesy of Private Collection.* $90-190.

"Deco Tel" Flag desk stand. American Telecommunications Corp. 11.5". *Courtesy of Private Collection*. $25-35.

Mickey Mouse telephone. American Telecommunications Corp. Disney Productions. 14.5" x 8.25". *Courtesy of Private Collection*. $75-150.

Piano telephone, the piano keys are the dials. Columbia Telecommunications Group, Inc. 3" x 7.25". *Courtesy of Private Collection*. $20-25.

Antique car telephone. Telemania, by Kash 'n Gold. Made in Taiwan. 3.5" x 10".
Courtesy of Private Collection. $35-55 ea.

Something every teacher should have, an apple telephone, 1985. Touchtone, 4". *Courtesy of Private Collection.* $20-25.

Canadian flag desk stand. Bell of Canada. 12" x 5". *Courtesy of Private Collection.* $20-25.

Purple Italian telephone; any ideas as to what it may be? Plastic, 6.75" x 8". *Courtesy of Private Collection.* Italian phone, $125-200.

Not just a telephone, but a toy as well. Lego, 1983. "Superblocks Telephone" by Tyco Industries Inc. Moorestown, NJ. 4" x 7" x 8". *Courtesy of Private Collection.* $35-60.

I am sure Alexander Graham Bell never pictured the telephone in this way! A Shoe Fashion Fone, 1987. Columbia Telecommunications Group, Inc. East Rockaway, NY. 5.75" x 9.5". *Courtesy of Private Collection.* $20-25.

Chapter Nine
Memorabilia

For the collector interested in telephony, collecting is not necessarily complete with a collection of telephones. There is a whole life of collecting which is related to the telephone, including parts, batteries, signs or advertising, some of which is shown in this chapter.

Battery jar from the National Carbon Co., Cleveland. *Courtesy of Private Collection.*

The Samson Battery No. 2, patented 1896. The carbon insert was made in France. Glass, 7.5" x 3.75" x 3.75". *Courtesy of Norman M. Mulvey.*

Glass battery jar, by ECW. 7.25" x 4". *Courtesy of Private Collection.*

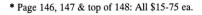

* Page 146, 147 & top of 148: All $15-75 ea.

Front & back views of The Samson Battery No. 2, patented Sept.
22, 1896. 7.5'' x 4'' x 4''. *Courtesy of Private Collection.*

Two glass battery jars, unmarked. Each 7.75'' high. *Courtesy of
Private Collection.*

Glass battery jar, Edison Manufacturing Co., Orange, NJ. B.S.C.O., Primary Battery No. 1, signed Thomas Edison. 12'' x 7'' diam. *Courtesy of Private Collection.*

Battery jar, made by Willard. 8.25'' x 2.75''. *Courtesy of Private Collection.*

Two types of Western Electric extension bells. Each 4.75'' x 5.75''. *Courtesy of Private Collection.*

* Page 149 & top left of pg.150: All $35-150 ea.

Sleigh bells used as extension bells. Western Electric. 4.75'' x 5.5'' x 4.75''. *Courtesy of Private Collection.*

Cowbells used as extension bells. Kellogg Switchboard & Supply Co., Chicago. 5.5'' x 6'' x 4.75''. *Courtesy of Private Collection.*

Large bells used as extension bells. 5.5'' x 8.25'' x 5''. *Courtesy of Private Collection.*

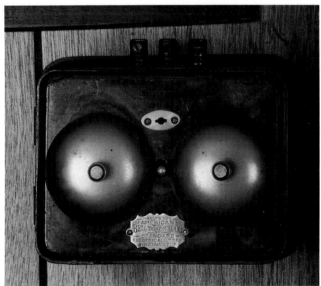

Extension bells, c. 1880s. Made by Standard Electric for the Bell Telephone Co. Burl wood, 4.25'' x 6'' x 5.75''. *Courtesy of Norman M. Mulvey.*

Group of extension bells, c. 1900s. Left to right: cow bells, sleigh bells, wooden bells, C.W. Williams, Jr. extension bells. Extension bells helped to distinguish which telephone was ringing. Also were placed in a room away from the telephone so that one knew when the phone rang even when not near the telephone. Avg. 5'' x 5.5'' x 5''. *Courtesy of Norman M. Mulvey.*

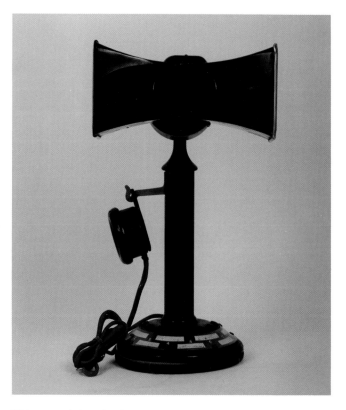

Extension bells, c. 1879. Made by Charles Williams, Jr. for the National Bell Telephone Co. Walnut, 5.75'' x 5'' x 5''. *Courtesy of Norman M. Mulvey.*

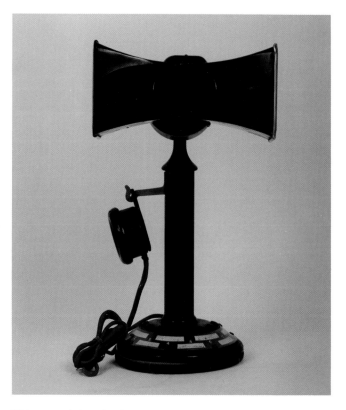

Whisper mouthpiece, dated Dec. 1921, attached to a Western Electric telephone. Mouthpiece by Hush-a-phone Corp., NY. *Courtesy of Norman M. Mulvey.* $100-200.

Unique sign in size and condition, c. 1907. Ingram-Richarson Mfg. Co., Beaver Falls, Pennsylvania. 8" x 5". *Courtesy of Norman M. Mulvey.* $425-550.

Do not disturb sign, 7" x 3.5". Warning sign, 12" x 4". *Courtesy of Dennis & Jeanne Weber.*

Public station sign. *Courtesy of Norman M. Mulvey.* $50-200.

Sign on both sides. Tin, 14" x 20". *Courtesy of Dennis & Jeanne Weber.* $50-200.

Porcelain enamel on steel, left: 11" x 12", right: 8.5" x 11". *Courtesy of Dennis & Jeanne Weber.* $50-200.

Telephone booth sign, c., 1910. New England Tel. & Tel. Co. 3" x 6.5". *Courtesy of Norman M. Mulvey.* $50-200.

General sign used by all Bell subsidy companies, c. 1890s. Made by the Imperial Enamel Co., Birmingham, England. 18" x 17". *Courtesy of Norman M. Mulvey.* $50-200.

Public station sign. AT&T, Western Union. Steel, 12" x 18". *Courtesy of Dennis & Jeanne Weber.* $50-200.

Fan showing a woman making a call. *Courtesy of Nick Zervos.*

Redburn drugstore calendar attachment opens up to a telephone list on one side, or a calendar on the other. *Courtesy of Private Collection.* $15-150.

* Page 153: All $25-75 ea.

Courtesy coin box attachment. 2.75'' x 5'' x 1.5''. *Courtesy of Norman M. Mulvey.*

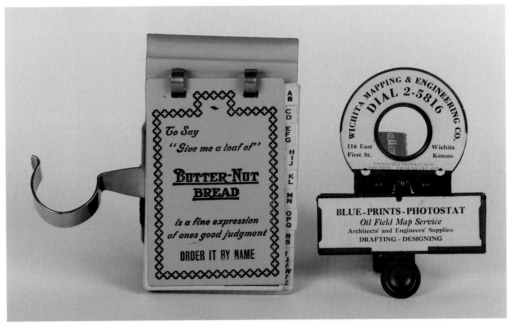

Butter-nut bread telephone book, 5'' x 5.25''. Wichita Mapping & Engineering Co. fold-up telephone book, 4.5'' x 3.25''. *Courtesy of Private Collection.*

Notepad attachments for telephones advertised all types of companies. This one displays Wm. J. McCuliff. 5.25'' x 3''. *Courtesy of Private Collection.*

Three attachments for the telephone, all with pencil attachments. Saniphone service, 1.25'' x 1.5''. Chas H. Case and Sierra Paper Co, Los Angeles. 1.75'' x 2''. *Courtesy of Private Collection.*

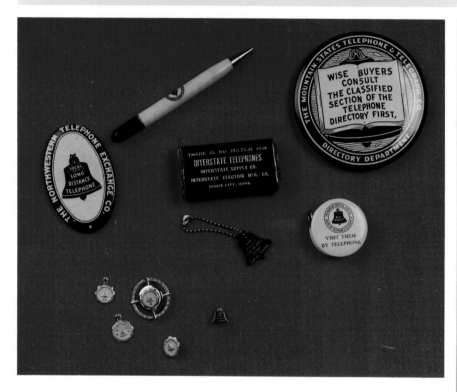

"Positive" Telephone Reminder advertising attachments, Christ E. Lamparter. 1.25" x 4.75". *Courtesy of Private Collection.* $25-75.

Various telephone memorabilia, including a paperweight mirror from "Mountain States Tel. & Tel. Co.," a tape measure, an "Interstate Telephones" matchbox, a key chain, a long distance pen, and jewelry and a small pin. *Courtesy of Larry Garnatz.* Service pins, $20-65 ea.

"Home" telephone index. Cardboard, 8" x 4". *Courtesy of Private Collection.* $25-75.

The Bell Company cafeteria place setting. *Courtesy of Barry Erlandson.*

Advertising fan and ashtray. *Courtesy of Larry Garnatz.*

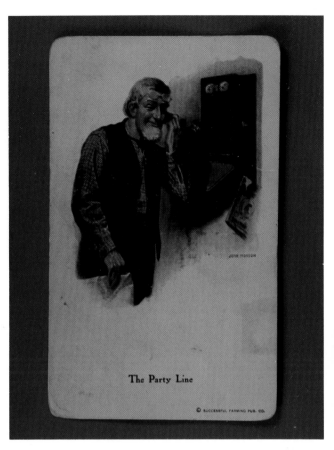

Post card, "The Party Line" by John Morton. Copyright by Successful Farming Publishing Co. *Courtesy of Larry Garnatz.*

Bell lampshades, used in telephone offices as hanging lampshades. Glass. *Courtesy of Private Collection.* $425-550.

Bell paperweights, c. 1910s-20s. The Missouri & Kansas Telephone Co. went out of business in 1918 and became part of Southwestern Bell. Glass, 3.5" x 3.5"; small weight, 2.25". *Courtesy of Dennis & Jeanne Weber.* $95-250.

Open and closed view of a decorative table which hides the telephone in the cabinet. 52'' x 23.5'' x 14''. *Courtesy of Norman M. Mulvey.*

Tri-state Telephone Company calendar, December, 1916. *Courtesy of Larry Garnatz.*

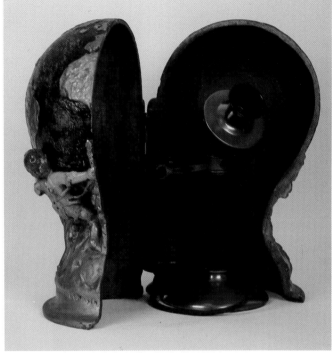

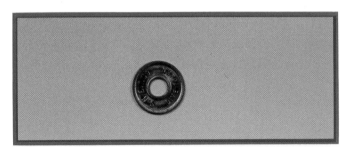

Telephone token, c. 1900. 75''. *Courtesy of Norman M. Mulvey.*

Open and closed view of the "Hide-A-Phone", c. 1920s. Desk stands were not considered to be decorative in the 1920s so they developed a way to hide them within a decorative piece. Open view shows a Connecticut Tel. & Electric Co. desk stand, Meriden, CT. Papier mâché, 13" x 8". *Courtesy of Norman M. Mulvey.* $350-700.

Chapter Ten

List of Companies

Acme Telephone and Mfg. Company

This telephone manufacturer was organized circa 1900-1901 in Chicago, Illinois, founded by W.O. Meissner.

American Cushman Telephone Company, Inc.

Sylvanus Cushman made and sold telephones in 1884 in the Midwest. On January 26, 1886, the company incorporated and was based in Chicago, Illinois. In 1888, they went out of business after court action for infringement.

American Speaking Telephone Company telephone with Edison transmitter, Edison spoon receiver, and a telegraph signaling system, c. 1879. Manufactured by Western Electric. Walnut, 8.5" x 8" x 7.5". *Courtesy of Norman M. Mulvey.* $1200-2000.

American Electric three box paddle phone, c. 1895. Hunnings transmitter. It was necessary to hold the paddle down to talk. Walnut, 30.5" x 13" x 8". *Courtesy of Norman M. Mulvey.* $1200-1800.

American Electric Telephone Company

American Electric was founded in 1894 in Kokomo, Indiana by P.C. Burns. In 1897, the company moved to Chicago and purchased the Keystone Mfg. Co. located in Pittsburgh.

American Speaking Telephone Company

The American Speaking Telephone Co. operated from 1877-1879 as a branch of the Western Union Telegraph Co. in New York City. Western Union opened this new branch for telephone operations to compete with Bell for rights to the telephone. In December, 1877, they acquired rights to Elisha Gray's patents. In May, 1878, control of American Speaking Telephone was given to Gold and Stock Telegraph Company, a subsidiary of Western Union. On November 10, 1879, business ceased due to court action for infringement of Bell's patents. In January, 1880, the company was annexed by the Bell Telephone Company.

American Toll Telephone Company

This company was formed in April, 1900 in Cleveland, Ohio by W.A. Foss and E.F. Kaneen as a manufacturer of telephones. Later the company was acquired by the Baird Manufacturing Company, and operated from 1901-1904.

Julius Andrae & Sons

The Andrae company was primarily an electrical parts supplier and jobber. They assembled and marketed a number of different telephones under their own name, but did not actually manufacture their own telephone equipment. The company was founded in 1893 in Milwaukee, Wisconsin by Julius Andrae and his son H.P. Andrae.

Automatic Electric upright desk stand, c. 1919-1940s. This type was made with and without a dial. Some were steel or iron, while the later models were bakelite. *Courtesy of Private Collection.* $100-175.

Automatic Electric Company

The company was formed in 1901 as an offshoot of the Strowger Automatic Telephone Exchange Company. Joseph Harris, the founder of the new Automatic Electric (A.E.) based in Chicago, had been an integral part in the start of the Strowger Co. A.E. at first manufactured apparatus under patents held by the parent Strowger Co. On March 27, 1908, A.E. acquired all the stock and patents of the Strowger Company. In 1955, A.E. was acquired by General Telephone and Electronics (GTE).

Baird Manufacturing Company

Edward P. Baird formed the Baird Mfg. Co. in the summer of 1900 in Chicago, Illinois. Between 1901 and 1905, the company acquired five companies including: American Toll Telephone Co.; Buffalo Automatic Mfg. Co.; Western Japanning Co.; Meyercord-Batterman Co.; and Norstrom Lock-Out & Telephone Mfg. Co.

Bell Telephone Company

On July 9, 1877, the Bell Telephone Company was formed with Charles Williams Jr.'s shop as the sole manufacturer for the Company. Theodore N. Vail became General Manager of the new Bell Telephone Company in June, 1878. In March, 1879, the Company became known as the National Bell Telephone Company with Colonel William H. Forbes as President.

In the fall of 1879, Western Union gave up all its patent claims and facilities to the Bell Company, giving the National Bell Company a legal monopoly of the telephone business that would last until the expiration of A.G. Bell's patents in 1893 and 1894. In March 1880, the company began to be called the American Bell Telephone Company.

In 1881, the American Bell Telephone Company bought Western Union's controlling interest in Western Electric. In 1885, a subsidiary of American Bell was formed to build and operate long distance lines under the name of American Telephone and Telegraph (AT&T) with Theodore N. Vail as president. In 1899, American Bell moved its headquarters from Boston to New York.

On March 27, 1900, AT&T took over American Bell and the child swallowed the parent.

Bell Telephone Company of Canada

Before the Bell Company of America was incorporated, Bell gave the Canadian patent rights to the telephone to his father, Melville Bell, who lived in Brantford, Ontario. His father traded 25% for the rights to Charles Williams of Boston in exchange for 1,000 telephones. The company was set up on April 29, 1880 by Melville Bell and Charles Sise.

James Cowherd of Brantford made the first "coffin-type" telephones for Bell Canada from 1878-1881. In 1882, Bell Canada formed a "Mechanical Department" which in 1895 became the Northern Electric Mfg. Co.

No. 4 Bell Telephone of Canada. IN the United States this same phone is a No. 10 and would have a black long pole receiver instead of a double red band. Patented 1899. *Courtesy of Art Hyde.* $1200-2000.

Canadian Machine Telephone Company. This very rare phone is located in the Ottawa, Ontario Science Center Museum. *Courtesy of Art Hyde.* Special.

B-R Electric & Telephone Manufacturing Company

Formed in 1903 in Kansas City, Missouri with F.M. Bernardin as President. The company formed as a result of the Kansas City Telephone Mfg. Co. and B-R Electric Co. merging. They manufactured a variety of wall sets and two distinct candlestick sets. They have a memorable mark on their telephones seen on two other types of candlesticks as well.

Canadian Machine Telephone Company

Formed in Brantford, Ontario.

Century Telephone Construction Company

Century was founded in Cleveland, Ohio in 1899. They moved to Buffalo, New York in 1902. In 1907 the company acquired the Williams-Abbott Electric Co. and continued to manufacture their line for an uncertain period of time. Sometime thereafter, Century changed its name to Federal Tel. & Tel. Co., though the name Century was used as late as 1913. Century is probably best known for the "split-shaft" tapered candlestick.

Chicago Telephone Supply Company

In 1869, the company was formed in Chicago, of course, and moved operations to Elkhart, Indiana in 1902. In 1911, the company began to manufacture "Briggs" auto magnetos. They also manufactured thousands of telephones for Montgomery Ward and Sears & Roebuck, many of which are easily found today.

Chisholm Dunn Telephone Appliances Ltd.

The factory was located on Colborne Street in Lohdon, Ontario. The company was in business until 1914 when it was apparently taken over by Bell of Canada.

Clark Automatic Telephone Switchboard Company

Founded in 1900, Providence, RI, Clark Automatic manufactured automatic intercoms.

Chicago Telephone Co., c. 1903-1905. Commonly called a Chicago "potbelly" due to the bulge in the stem. *Courtesy of Art Hyde.* $200-300.

Chisholm Dunn Telephone Appliances Ltd., patented in United States and Canada in 1911. Very rare telephone. *Courtesy of Art Hyde.* $300-450.

Century Telephone Construction Co., "split shaft," c. 1905-1907. The two sides split in half when unscrewed. Pictured together and split. 10.75". *Courtesy of Private Collection.* $400-600.

Columbia Telephone Manufacturing Company

This company was located in Chicago, Illinois and founded in 1901. They manufactured multiple exchange and toll line switchboards, magneto bells, transmitters, receivers and various other telephone-related items. In 1901, James G. Nolan was the company engineer and held the patents on most of the telephone equipment manufactured.

The Columbia Telephone Manufacturing Co.

Located in New York, New York, advertising for the company's telephones began as early as December, 1894. Patents were held under Jas. W. McDonough and H.H. Eldred with uniquely designed transmitters and receivers, very collectible today.

Connecticut Telephone and Electric Company

Based in Meriden, Connecticut, the company often purchased parts from other large manufacturers and added their own parts.

50-station wall intercom with two links, Auto-phone, (hung in a warehouse) c. 1910. Made by S.H. Couch, Inc., Boston. Receiver marked Norfolk Downs, Mass. *Courtesy of Private Collection.* $100-200.

S.H. Couch, Inc.

Based in Boston and Norfolk Downs.

Cracraft-Leich Electric Company

In July, 1907 in Genoa, Illinois, J.P. Cracraft and O.M. Leich formed the Cracraft-Leich Electric Company. In August they acquired the Eureka Electric Company and in 1916 the company became simply Leich Electric.

Davis and Watts

In 1878, A.G. Davis and H.C. Watts formed one of the first companies to manufacture telephones for the Bell Telephone Co. The company, located in Baltimore, Maryland was contracted under license to make telephones for Bell. Once the Bell Co. made agreement with Western Union to acquire the Western Electric Mfg. Co., Bell canceled its contracts with Davis and Watts. Davis and Watts was dissolved and annexed to Viaduct Mfg. Company of Baltimore, in June, 1883. Davis and Watts were known particularly for the "coffin" telephones they produced for Bell.

Dean Electric, Elyria, Ohio, c. 1903-1914. *Courtesy of Art Hyde.* $150-250.

Dean Electric Company

Dean Electric was formed in December, 1903 in Elyria, Ohio. The company acquired Eastern Telephone Mfg. Company, a company which had obtained Sun Electric Mfg. Co. some time earlier. In 1914, Dean became the Garford Mfg. Company and in 1916 they were acquired by Stromberg-Carlson.

DeVeau single station upright desk stand intercom, c. 1915. Transmitter marked Stanley N. Paterson, DeVeau. 11.25". *Courtesy of Private Collection.* $150-250.

produced the same line of products as they had under the previous name, with the trade name "Samson" identified on their line of intercoms, for which they were well known. They are believed to have used considerable equipment manufactured by other companies and perhaps did not even manufacture any of their own.

Elston Telephone Company

Based in Muscoda, Wisconsin.

Enochs Electric Company

Founded in St. Paul, Minnesota by Claude D. Enochs in 1908. The company was a distributor

DeVeau Telephone Manufacturing Company

Based in Brooklyn, New York.

Electric Gas Lighting Company (EGL Co.)

Based in Boston, c. 1880, Electric Gas is probably known for distributing the Samson glass battery jar. The initials are embossed on the glass jar as E.G.L. Co. They also manufactured and assembled intercoms, and specialized hotel signal boards, or annuciators, with names such as Rival, Ross, Noxall, Perfect, and Dandy. Electric Gas seems to be the sole manufacturer of the Samson intercom, referred to as their line of Samson Junior telephones. The company changed its name to Electric Goods Mfg. Co. (EGM Co.) in 1904, though some products used the early name up until 1905.

Electric Goods Manufacturing Company

In 1904, the Electric Gas Lighting Company became the Electric Goods Mfg. Co. They

Elston Telephone Company with Mae West backboard. Oak. *Courtesy of Art Hyde.* $250-400.

and a manufacturer. Claude Enochs, founder and president evidently had over 10 years of telephone experience before organizing the company, and held patents on both a transmitter and a receiver.

L.M. Ericsson & Co., c. 1895. *Courtesy of Private Collection.* $1500-2500.

L.M. Ericsson Manufacturing Company

A European company founded in Stockholm, Sweden on April, 1876 by Lars Magnus Ericsson as a manufacturer and repair company for telephones and other electrical instruments. In 1878, they began to manufacture their own telephones. In 1897, Ericsson established a factory in St. Petersburg, Russia. They began selling telephones in the United States through agents in the 1890s. Couch and Seeley were one such agent for the company. A sales office was opened in New York in 1902, and, in 1905, Ericsson opened a factory in Buffalo, New York which began production in 1907.

Fearing a diminishing supply of wood, the company decided to manufacture an all-steel fiddleback telephone which was manufactured in Buffalo and introduced in 1907. Several early candlesticks are quite desirable among collectors as well as the cradle desk set with the micro handpiece and a Stockholm manufactured desk known as the "Eiffel Tower."

Eureka Electric Company

The company either began in 1896 or 1898 in Chicago. In 1903 the company moved to Genoa, Illinois, and was purchased by Cracraft-Leich in 1907.

Farr Telephone & Construction Company

This manufacturer was founded in 1895 in Chicago by Charles Farr, N.B. Farr and E.W. Hurst.

Federal Telephone & Telegraph Company

Originally named Century Telephone Construction Co., they changed their name to Federal and moved to Buffalo, NY. In 1907, the company was purchased by the Williams-Abbott Electric Co., Cleveland, Ohio but remained in Buffalo. In 1920 they were bought by Corwin Electric Co. In 1956, Federal was sold to International Telephone & Telegraph (ITT).

Fisk-Newhall Telephone Mfg. Company

Another Chicago manufacturer founded in 1902 by Henry M. Fisk and Simeon F. Newhall. They manufactured a complete line of telephones.

Garford Mfg. Company

Originally the Dean Electric Company, it was named Garford in April, 1914. The company was based in Elyria, Ohio and in 1916 was purchased by Stromberg-Carlson.

Gilliland Electric Mfg. Company

Ezra T. Gilliland (1847-1903) was a designer and manufacturer of telephones and switchboards for the Indianapolis Telephone Co. in 1879. Gilliland Electric was one of the original manufacturers the Bell Telephone Co. under the Bell license until 1882, when they were acquired by Western Electric.

Globe Automatic Telephone Company

Formed in 1900 in Chicago, the Globe Co. manufactured a line of automatic intercoms and central office exchange equipment. They purchased the National Automatic Telephone Co. in 1901. The company apparently had trouble competing with the introduction of the Strowger Dial system and was dissolved in 1911.

Gold and Stock Telegraph Company

The company was formed in 1873 as a branch of Western Union and entered the telephone business in May, 1878, when it acquired control and management of American Speaking Telephone Co. The company was ideal as an agent for American Speaking Company's central office exchanges since it owned and operated a vast network of telegraph lines. In September, 1877, the company had acquired rights to Dolbear's patents. November, 1879 they ceased all telephone operations so as not to infringe on Bell's patents. In 1880, the Gold and Stock Telegraph Co. was purchased by Metropolitan Telephone Company.

Green Telephone & Electric Mfg. Company

Not much information has been found about this company. They seemed to have formed c. 1902 in Milwaukee, Wisconsin. Evidently the company was an assembler, not a manufacturer of telephones and often used Julius Andrae parts.

Gray Telephone and Pay Station Company

William Gray, received a patent on August 13, 1889 for a coin operated device to work the telephone. He formed the Gray Telephone and Pay Station Company which held a monopoly on the pay station, well into the 20th century. It was acquired by the Automatic Electric Company in 1948.

J.R. Holcomb and Company

The J.R. Holcomb company was formed in 1872 when Mr. Holcomb located in Mallet Creek, Ohio. It was in 1878, however, that Holcomb, as one of the first telephone manufacturers and inventors, introduced an "automatic telephone" also known as an "acoustic," "string," "tight-wire," or "mechanical" telephone. It was guaranteed to "work for one mile" and cost "only $4.00 per set." The company sold over 10,000 sets. A popular collector's item is a square "string phone" manufactured in 1881 by the Holcomb Co. They are often found in mint condition.

Holtzer-Cabot Telephone Company

The company was formed in 1889 by Charles Holtzer who had acquired the stock and machinery from the Charles Williams, Jr. Co.,

Gray Telephone Pay Station. *Courtesy of Burt Prall.* $500-600.

the machine shop of the first telephone. A Mr. Cabot sold his interests in the company in 1892.

Illinois Electric Company

The company was formed in Chicago in 1898. They advertised that they manufactured their own line of telephone equipment, however, it is believed that they were largely a distributor.

International

Based in Chicago. Telephone parts for any International telephone are hard to find.

Interstate Electric and Mfg. Company

The firm, founded c. 1906 in Sioux City, Iowa, specialized in the manufacture of transmitters, ringers, magnetos and common battery wall sets.

Kellogg Switchboard & Supply Co.

Milo G. Kellogg ventured into the telephone business at an opportune time, planning to enter the market as the Bell patents on telephone design and manufacture would expire. Kellogg worked on improvements on the telephone for years so that when he did open his business, he was ahead of most.

The company was formed in 1897 in Chicago, in Elisha Gray's old workshop to be exact. The company was then bought by Western Electric whicn Milo Kellogg left the company for a few years due to ill health in 1902. The corporation was returned to Kellogg in 1909 by the courts of Illinois. Kellogg made all of its own parts and was a large manufacturer of switchboards, candlesticks and wall sets. They developed a switchboard with harmonic ringing, reverse type transmitter and phenol parts such as receiver shells and mouthpieces.

Kellogg surrounded himself with engineers such as Francis W. Dunbar, William W. Dean, Kempster B. Miller and Franz J. Dommerque, all known as pioneers who contributed to the origins of independent telephony. In 1952 the company was purchased by ITT.

Keystone Telephone Co.

The Keystone Company apparently began in 1894 with a factory in Pittsburgh assembling and manufacturing two and three box telephones. These telephones usually carried a Keystone logo-shaped tag with the company name. In 1898, the company merged with the Northwestern Telephone Equipment Co. and the American Electric Co. under P.C. Burns, owner of American Electric.

A model of the Western Telephone Construction Co. two box and Manhattan Electric Supply Co. three box telephone is known to have been marketed with the Keystone, Pittsburgh name tag. Some advertisements from the 1901-1905 era use the name Keystone Electric Telephone Co., Pittsburgh. One speculates that American Electric probably retained a sales outlet in Pittsburgh under the name Keystone Electric to sell equipment to the independent companies.

The Keystone Telephone Co. of Philadelphia (supposedly related to the original company) opened its business circa 1900. American Electric was evidently the sole manufacturer of Keystone, Philadelphia telephones, most of which are well-marked, while the candlesticks and wall sets have the Keystone logo cut into the switch hooks. Keystone of Philadelphia competed with the Bell system until absorbed by Bell in 1945-46. The city of Philadelphia, however, retained the Keystone telephone lines which served the fire, police, school, and bus services and are still in use today.

Leich Electric Co., Genoa, Il, c. 1915. Marked receiver 11". *Courtesy of Private Collection.* $150-225.

Leich Electric Co.

Originally called the Cracraft-Leich Co., the name changed in December, 1916 to Leich Electric. Manufacturers of telephones.

Lincoln Telephone & Telegraph Co.

A telephone company and assembler founded in 1903 in Lincoln, Nebraska.

Lockwood

Long Distance Telephone Mfg. Co.

This manufacturer of telephones was founded in 1905 in Chicago. The company moved to South Bend, Indiana in 1907. Evidently, another company under the similar name of The Long Distance Telephone Sales Company was also organized in 1905 but is not believed to have been related.

Manhattan Electrical Supply Co., Inc.

The company was incorporated in Manhattan, New York on September 10, 1889.

Maryland Telephone Mfg. Company

The company factory was located in Baltimore, Maryland while the sales office was in New York. They opened in 1898 and the factory burned down in 1904.

McLean Telephone Company

Based in Greencastle, Indiana.

McLean Telephone Company telephone. *Courtesy of Art Hyde.* $400-600.

Manhattan Electric long distance telephone, c. 1900. Split battery box. Walnut. *Courtesy of Art Hyde.* $750-1500.

Monarch Telephone Mfg. Company

In June, 1901 the Monarch Telephone Mfg. Company was founded in Chicago, Illinois. In 1925-1926, the company was purchased by American Electric which then became Automatic Electric.

Montgomery Ward and Company

Some time around 1900, Montgomery Ward began selling telephone products manufactured by the Chicago Telephone Supply Company. A brass nametag was attached to most wall sets and a very few marked transmitters are known to have been manufactured, as well as a number of magnetos, carrying a Ward's nameplate.

Munsen Automatic Telephone Company

In 1907 Augustus Munson invented and operated a telephone exchange that utilized automatic switching with a 100 number dial. Approximately 100 telephones of this type were manufactured making them highly collectible. The Company was located in North St. Paul, Minnesota. In 1912 the company sold out to Northwestern Telephone Exchange Company.

Murdock & Taber

The company was formed in the early 1890s as a manufacturer. One of the founders, Wm. J. Murdock, is probably better known for having developed and manufactured the well-known SOLID receiver which was widely used by many early New England independent companies. SOLID receivers typify a much sought-after collector's item. In 1904, Taber was dropped from the company name leaving it to be called the Wm. J. Murdock Co. in Chelsea, Massachusetts.

National Automatic Telephone Company

Founded in 1897 in Salina, Kansas, the company was both manufacturer and assembler. They used early Chicago parts but had their own unique hook switch design. They used a simple dial unit which could not compete with the Strowger dial. They sold out to Globe Automatic in 1901 who in turn went out of business in 1911. Only three complete National Automatic telephones are known to exist, and all are wall phones.

North Electric Company

The North Electric Company was founded in 1886 and incorporated in 1889 in Cleveland, Ohio. In 1918, after having acquired Telephone

Toploader, c. 1905. North Electric Co., with triplet assembly, oak. *Courtesy of Norman M. Mulvey.* $600-750.

Improvement Company in 1912 and moving to Galion, Ohio, the company changed its name to North Electric Manufacturing Company. North introduced the following items: Type F "potbelly" candlestick, 1900; Type E "toploader," 1901; No. 1 pulse sending dial used by Kellogg, Stromberg and Automatic Electric, 1922.

Northern Electric & Manufacturing Company

Northern Electric became the Canadian manufacturing arm to Bell Canada in 1895 as Western Electric was the manufacturer to the Bell Telephone Company in the United States. It was formed from a dominion charter granted by the government to allow the forming of a separate company. Telephone equipment was manufactured specifically for use by Bell Canada. In 1914, the name was shortened to Northern Electric Company.

Niles Chair Company

Based in Niles, Michigan, the company made string telephones in the 1880s.

Phoenix Electric Telephone Company

Phoenix, established in 1895 in New York, manufactured a variety of telephones, including some very interesting two and three box sets. The equipment made by Phoenix was quite attractive and of good quality and is therefore much sought-after by collectors today.

Premier Electric Company

The Premier Electric Company, founded in 1897, Chicago, Illinois, primarily manufactured wooden magneto wall sets. They went out of business in 1933.

Rawson Electric Co.

Founded in 1894 in Elyria, Ohio, the company was sold to Dean Electric in 1903 which then became Garford Mfg. Co. in 1914. Garford was sold to Stromberg-Carlson in 1916.

Screw Machine Products Company

Founded in 1916 by Alton E. Stevens in Providence, RI. The company is best known for its Select-O-Phone line of intercom equipment. The sets were probably their own but the transmitters on the wall sets and candlesticks were made by Stromberg-Carlson. The receivers were from Kellogg. Though the candlestick has a perch that looks like a Kellogg, it is not adjustable like a Kellogg. Candlesticks are usually either mostly brass or steel and the wall sets are seldom seen in wood.

Sears Roebuck & Company

The Sears catalog was advertising telephones as early as 1897. They were distributors, mainly for products manufactured by the Chicago Telephone Supply Company. A brass nametag with the Sears name was attached to most wall sets. Many of the telephones were made of iron or steel therefore these phones are quite easy to recognize as they often rust after time. Also, nickel plating does not adhere well to iron which led to the extensive practice of painting many of the nonbrass parts on the Sears products with black paint.

Select-A-Phone Company of America

Located in Denver, Colorado, this company should not be confused with the Screw Machine Products Company which produced the Select-O-Phone. The Select-A-Phone manufactured by the company was advertised as providing private service on party lines in the January, 1921 issue of *Telephony* magazine.

Selectaphone Manufacturing Company

Another selectaphone company, not to be confused with the Select-O-Phone manufactured by The Screw Machine Products Company or the Select-A-Phone Company of America.

The Selectaphone was advertised in 1902 in *Telephony* magazine as a device developed for use on party lines which operated through a unique step-by-step switching machine. The subscriber end was equipped with a unit built into a wooden box which was installed between the incoming line and any telephone set.

The system allowed "central" to selectively ring any telephone without disturbing the others. It also provided complete privacy to the talking parties. A busy signal would display on each Selectaphone when the line was in use. Such a system is known to have been installed for the Belmont County Telephone Company of Ohio in July, 1902. The line was 14 miles long and had 10 stations on it.

Shawk and Barton

Shawk and Barton founded the company, January, 1869 in Cleveland, Ohio which in April became Gray & Barton. In April, 1872, the name changed to Western Electric Manufacturing Company and it was consolidated with Western Union. The consolidation with Western Union gave them rights to Gray's patents (receiver), Western Union's rights to Thomas Edison's carbon transmitter, American Speaking Telephone Co.'s agent, Gold & Stock, which controlled Dolbear's patents rights and Phelps receivers. They manufactured telephones and equipment for Western Union subsidiaries.

In 1881, the name changed to the Western

Electric Company and American Bell Telephone bought Jay Gould's interests from Western Union. Western Electric became the manufacturer for American Bell.

Siemans & Halske

Based in Berlin, the company also had offices in Buffalo, NY. One of the first manufacturers of "field phones" produced for military use.

Standard Electrical Works

Standard Electrical Works began in 1878 with offices in Cincinnati, Ohio. The company was contracted to make telephone instruments and switchboards for Bell the next year. They also manufactured magneto bells for Post & Co. in 1882. They transferred controlling amount of stock to Western Electric Company in 1884 and in 1888 the name changed to Standard Electrical Co., Inc. In 1902, the name was Standard Electric Co. and in 1907, the company became a branch house of Western Electric as Standard Electric.

Standard Telephone Manufacturing Company

The company started as Drake Telephone Company in McConnelsville, Ohio. In 1905, known as the Knox-Dickey Telephone Manufacturing Company, they made central energy equipment. In 1907, the company purchased Williams Telephone & Supply Co. from Cleveland and moved the plant to Portsmouth, Ohio adding magneto apparatus to their line under the name of Standard Telephone Manufacturing Company.

Sterling Electric Company

The company began in Chicago in 1898 and moved the following year to LaFayette, Indiana. Sterling manufactured everything for their telephone except hard rubber, porcelain and glass items. In 1911, LaFayette Electric & Manufacturing Co. bought Sterling Electric.

Stromberg-Carlson Telephone Manufacturing Company

Alfred Stromberg and Androv Carlson went into partnership in 1894 when they both invested $500 in a new company. This was the start of the Stromberg-Carlson Telephone Manufacturing Company, formed in Chicago, Illinois. In 1903, the company moved to Rochester, New York and incorporated. In 1905, both Stromberg and Carlson left the business and returned to Chicago.

The "oil can" candlestick is probably the best known Stromberg-Carlson telephone made from c. 1898-1905. They also produced an elaborate desk vanity which is a choice collector's item today.

Strowger Automatic Company

It all began in 1891 with Almon B. Strowger. The Kansas City undertaker felt deceived because his competitor's wife worked the local switchboard. He feared she was deliberately directing his calls to his competition, so he determined to develop a telephone system which would eliminate the go-between switchboard operator. In doing so, he developed the first automatic dial system known as the Strowger dial.

Fortunately for Strowger, his invention entered the scene at the height of activity for novelty items. In these times, enterprising men such as Joseph Harris planned their fortunes on other's new ideas. Harris was a traveling salesmen who was waiting for some novelty to appear when he met up with Walter Strowger, Almon's nephew, and it all took off from there. Harris was a leading force in the foundation of the Strowger Automatic Telephone Exchange Company in 1891 taking its roots in Chicago, Illinois.

In 1901, the Strowger Automatic Company merged into the Automatic Electric Company with Joseph Harris as president. It remained Automatic Electric until 1955 when the company was purchased by General Telephone and Electronics.

Sumter Telephone Manufacturing Company

The company was formed in 1899 in Sumter, SC as a manufacturer of telephones. The Sumter Telephone Supply Co. was formed in 1913 to handle the business of the telephone department and to represent the Dean Electric Company. In 1920, Sumter Manufacturing was purchased by Splitdorf Magneto Company of New Jersey and was closed.

Sun Electric Manufacturing Company

Sun Electric was founded in 1900 in Philadelphia, Pennsylvania. In 1902, the company was sold to Eastern Telephone Company which was later acquired by Dean Manufacturing Co.

Sun Electric "fiddleback," c. 1902. Glass in front, oak. *Courtesy of Norman M. Mulvey.* $1000-1500.

Swedish-American Telephone Company

This Chicago-based company was founded in December, 1899 and chose its name as a marketing technique as their telephones were sold in the mid-West where a majority of customers were of Scandinavian descent. The company manufactured some of their own parts but acquired many of them from other manufacturers, such as L.M. Ericsson. The short "potbelly" candlestick introduced in 1900 is an example of a good collector's item. Swedish-American manufactured a wide range of telephone equipment. S-A was the parent company of the Waterford, Ontario company in Canada-Dominion Telephone Company. Circa 1920, the company ceased business permanently.

Taber & Mayer

Not much is known about this company, although they were known to have manufactured telephones in the Boston area circa 1898.

Universal High Power Telephone Company

This Seattle manufacturer and assembler produced a very large transmitter with intentions of improved transmission. The transmitter was mounted on a Leich Electric upright desk stand and was patented in 1917.

Utica Fire Alarm Telegraph Company

The company was formed in 1876 as a manufacturer of telegraph instruments and other such patented inventions of A.H. Palmer, one of the founders. With the invention of the telephone, the company manufactured telephones, switchboards, acoustic telephones, exchange apparatus, and other telephone related items.

Viaduct Manufacturing Company

The Viaduct Company was formed in 1883 from the dissolved Davis and Watts Company. The manufacturing plant was located in Elk Ridge, Maryland adjacent to the B & O Railroad Line. A brick viaduct was built to support the railroad tracks over a ravine next to the plant and this was the source of the company name.

Viaduct manufactured a line of unique and interesting telephones, much desired by the avid collector. An example is the "paddle phone." Viaduct also manufactured parts, especially top boxes, which were often used on the telephones of other companies. Their magnetos were similar to those used by Manhattan and others.

In 1894 a fire destroyed the plant in Elk Ridge but it was reconstructed shortly thereafter. In 1983 the name was changed to the Davis and Hemple Screw Machine Company.

Vought-Berger Company

Formed in 1900 in LaCrosse, Wisconsin by M.I. Vought and Marcy Berger. Although the company is best known by most for its "pendant" hanging telephone, the company made a full line of wall and desk stand telephones, switchboards and related items. They were eventually absorbed by Automatic Electric Company.

Wesco Supply Company

Wesco apparently was a distributor of telephones and supplies in St. Louis, Missouri. The company was formed from acquisition of the Western Electrical Supply Company in August, 1903.

Western Electric Company

In 1869, Elisha Gray and Enos M. Barton formed a partnership named Gray and Barton, establishing a manufacturing plant in Cleveland, Ohio. They produced electrical products consisting chiefly of telegraph instruments, electric bells, signal boxes, batteries and fire alarms. In 1877, they moved the operation to Chicago and became the Western Electric Company. Western Electric supplied Western Union with its telephone equipment from 1878-1879. In 1881, the Bell Company acquired Western Union's interest in Western Electric and in 1882, the company soon became the sole supplier of Bell equipment and a part of the Bell System. Enos Barton directed Western Electric for 40 years. A factory was constructed in Belgium in 1882, and in England, Germany, France, and Japan a little later.

Western Electric eventually divested themselves of many activities including the selling of their foreign factories to International Tel. & Tel. Company (ITT). They also sold their "jobbing" business to the employees in it, who organized a new corporation called the Graybar Electric Co. in tribute to the founders Gray and Barton. Graybar became the marketing arm for Western Electric.

In 1925, the engineering department of Western Electric was incorporated as Bell Laboratories, jointly owned by Western Electric and AT&T.

Fairly rare Vought Berger two boxers, c. early 1900s. Oak. *Courtesy of Art Hyde.* $600-800.

Western Electrical Company

This company, based in Omaha, Nebraska and founded in 1889 is believed to have distributed telephones manufactured by others as their company name is occasionally found on a transmitter faceplate name tag. In 1909 the name was changed to Johnson Electric Company.

Western Union Telegraph Company

Western Union entered the telephone industry in 1877 after having refused an offer to buy Bell's patents in 1876. Western Union used Edison, Dolbear, Gray and Phelps equipment to compete with the Bell Company until 1879 when they ceased all telephone operations after infringing on Bell's patents.

Wilhelm Telephone Mfg. Company

Founded in Buffalo, New York, November, 1898 as a manufacturer of telephones. A factory was opened in Dunnville, Ontario, Canada between September, 1907 and April, 1908. In 1913 the company ceased operations.

Williams-Abbott Electric Company

Founded in 1895 by Joseph A. Williams in Cleveland, Ohio. In 1897, Williams sold his interest in the company and formed the Williams Electric Company, which in turn became the Williams Telephone & Supply Company in 1903. Williams-Abbott Electric Company was sold in 1907 to Federal Telephone & Telegraph Company.

Williams Electric Co.

In 1897, Joseph A. Williams, after founding and selling his interest in the Williams-Abbott Electric Company, organized the Williams Electric Company, also based in Cleveland, Ohio. In 1903, Williams Electric became Williams Telephone & Supply Company.

Williams Telephone and Supply Company

Successor to the Williams Electric Company, founded in 1897 by Joseph A. Williams. The name became Williams Telephone and Supply Company in October, 1903 when Joseph Williams seems to have left the business. In 1907, Williams Telephone & Supply was acquired by Standard Telephone Mfg. Company of Cleveland.

Chapter Eleven

Sources for Collectors

International Telephone Collectors Clubs
Telephone Collectors International (TCI)
19 Cherry Drive, N. Oswego, IL 60543

American Telephone Collectors Association
(ATCA) P.O. Box 94 Abilene, Kansas 67410

Telecommunications Heritage Group
P.O. Box 499, Bishopbriggs Glasgow G64 3JR,
Scotland

Australasian Telephone Collectors Society
P.O. Box 566 Lane Cove, N.S.W. 2066, Australia

Australian Historic Telephony Society, Inc.
P.O. Box 194 Croydon, Victoria 3136, Australia

New Zealand Historic Telecommunications
Collectors Club
c/o Bevan Healey
71 Wycliffe Street, Onekawa Napier, New
Zealand

The Telephone Pioneers of America
The Telephone Pioneers of America is the world's largest voluntary association of industrial employees with over 740,000 pioneers in the 105 chapters and 1500 councils and clubs. The organization is made up of both men and women from the United States and Canada who have served for 17 years or more in the telephone industry.

The organization was formed by the actual "pioneers" who experienced the days of Alexander Bell. Today's pioneers maintain the strong feeling for the history and traditions of the business. It began as a social group, but in 1958 during the general assembly meeting of the Pioneers, "Community Service" was projected as an official part of the pioneer program, yet in many chapters these services had been in action since WW II. Today the pioneers have developed thousands of projects which they both fund and serve to help the handicapped.

The three sides of the triangular insignia of the pioneers represent the principal objectives of the group: fellowship, loyalty and service. The number on the bell at the center is the U.S. patent number for Alexander Graham Bell's patent. The date of 1875 commemorates June 2, 1875, when Bell verified his theory of the electrical transmission of speech. November 2, 1911 honors the date the organization was founded.

Bibliography

Boettinger, H.M. *The Telephone Book*. Croton-on-Hudson, New York: Riverwood Publishers Limited, 1977.

Brooks, John. *Telephone, The First Hundred Years*. New York: Harper & Row, 1976.

Contributions from the Museum of History and Technology. United States National Museum Bulletin 228. Paper 29: Development of Electrical Technology in the 19th Century: II. Smithsonian, Washington, DC, 1962. Pages 315-332.

Dommers, John J. *The Telephone Connection*. A Guide to the Identification of Old Telephones and Related Items. Madison, CT: Sachem Press, 1983.

Historical Fact Sheets of early Telephone Industry Related Companies. The Antique Telephone Collectors Association Publication.

Knappen, R.H. *History and Identification of Old Telephones*. Vol. 1 & 2. Compiled by R.H. Knappen, 1978.

Kunath, Richard. *Photograph Collection of Early Telephones and Equipment used by the Bell System, 1877-1904*. Compiled by Richard Kunath.

Preece, William Henry F.R.S. and Julius Maier, PhD. *The Telephone*. London: Whittaker & Co., Paternoster Square. New York: D. Van Nostrand Company, 1889. (Smithsonian Library).

Pool, Ithiel de Sola, ed. *The Social Impact of the Telephone*. Cambridge, Mass.: The MIT Press, 1977.

The Popular Science Monthly, Aug 1883. "The Telephone, with a Sketch of its Inventor, Philipp Reis. Channing, William F., M.D.

Plush, S.M. "Edison's carbon Telephone Transmitter, and the Speaking Phonograph." *Journal of the Franklin Institute* April, 1878. Vol. CV No.4.

Watson, Thomas A. *The Birth and Babyhood of the Telephone* An address delivered before the Third Annual Convention of the Telephone Pioneers of America at Chicago, October 17, 1913. American Telephone and Telegraph Company, 1929.

Telephone Collecting: Seven Decades of Design Kate E. Dooner. Here, in text and over 250 color photographs, a history of the design of the telephone is presented from Art Deco years to novelty phones of the 1980s. The largest telephone companies are discussed, including Western Electric, Automatic Electric, Stromberg-Carlson, Kellogg, and North Electric.

Size: 8 1/2" x 11"	Price Guide	128 pp.
ISBN: 0-88740-489-8	soft cover	$24.95

One Hundred Years of Bell Telephone Richard D. Mountjoy. From the Coffin sets of the 1870s to the Princess phones of the 1960s and beyond, this book explores the technology and the history of the telephone. This definitive work provides detailed information which will help identify a piece and will take the guess work out of dating equipment. For those who are restoring a telephone and would like to ensure its historical accuracy, this book will make it easy to match pieces correctly. 350 color photos
Size: 8 1/2" x 11"   Price Guide   176pp.
ISBN: 0-88740-872-9   soft

America's Oak Furniture Nancy Sc̲h̲... topher Biondi. The mellow tones ... become popular accents in a growi̲n̲... America today. With hundreds of ex̲a̲... graphs, the book is arranged by types ... tables. Hundreds of chairs are shown ... styles that were made. Famous maker̲s̲... more are well represented.
Size: 8 1/2" x 11"   Over 3(...
ISBN: 0-88740-158-9   soft

The Zenith TRANS-OCEANIC, Th̲... Bryant, AIA and Harold N. Cones, ... untold story of the Zenith Trans-Ocea̲n̲... and expensive series of portable radi̲... corporate archives and their own expe̲... writers, Bryant and Cones present ... Trans-Oceanic throughout its forty ye̲... of photos, documents and informatio̲n̲... radios, their collection, preservation ...
Size: 8 1/2" x 11"   Valu̲...
180 photo illustrations, many in col̲o̲...
ISBN: 0-88740-708-0   so̲...

Zenith Radio: The Early Years 1̲... and John Bryant, AIA with Ma̲r̲... Wade. Tells the fascinating story o̲f̲... history. Never before published ... information, as well as color portra̲... era. Complimenting the story is a ... every Chicago Radio Laboratory a̲r̲... by the company.
Size: 8 1/2" x 11"   Val̲...
ISBN: 0-7643-0367-8   so̲...

Philco Radio: 1928-1942 Ron Ramirez, with Michael Prosise A superb reference book on Philco, the leading radio manufacturer during radio's "Golden Age." Specifications for each model given. A year-by-year look at Philco's radio line, so that the reader may see what Philco had to offer each year between 1928, when the company began to make radios, and 1942, when World War II put a halt to radio production.

Size: 8 1/2" x 11"	Price guide	160 pp.
464 color, 87 b/w photos		
277 line drawings	soft cover	
ISBN: 0-88740-547-9		$29.95

Radios by Hallicrafters Chuck Dachis. This book includes over 1000 photographs of radio receivers, transmitters, and speakers, early television sets, electronics accessories and advertising material produced by this Chicago-based firm. Technical descriptions are provided for every known Hallicrafters model, including dates of production, model numbers, accompanying pieces, and

	Price Guide	225 pp.
	soft cover	$29.95

... **Coin Machines** Richard M. ... thorough look at slot machines, ... pinball games, trade stimulators, ... ated scales, all described in detail ... color. The text presents methods ... of machines, their history and ... on where to look for them.
...uide   192pp.
...rd cover $39.95

...strated Compendium 1877-1929 ...F. Paul. This thorough compen- ...tographs showing an incredible ...as internal-horn talking machines. ...e guide complement a wealth of ...complete work on antique talking ...anced collector.
...0 color photos   256 pp.
...rd cover $69.95

...D. Ball. This book is filled with ...ions, products, giveaways, sales ..., uniforms, and equipment of the ...storical black and white photo- ...s, as well as early airplanes and ...to fleet a̲r̲... ...s concise ...   Price ...hite ...   Soft C̲...